T0179648

Developments in Strategic Materials and Computational Design III

Developments in Strategic Materials and Computational Design III

A Collection of Papers Presented at the 36th International Conference on Advanced Ceramics and Composites January 22–27, 2012 Daytona Beach, Florida

Edited by
Waltraud M. Kriven
Andrew L. Gyekenyesi
Gunnar Westin
Jingyang Wang

Volume Editors
Michael Halbig
Sanjay Mathur

A John Wiley & Sons, Inc., Publication

Published by John Wiley & Sons, Inc., Hoboken, New Jersey.
Published simultaneously in Canada.

Library of Congress Cataloging-in-Publication Data is available.

ISBN: 978-1-118-20600-3
ISSN: 0196-6219

Printed in the United States of America.

10 9 8 7 6 5 4 3 2 1

Contents

ADVANCED MATERIALS AND PROCESSING FOR PHOTONICS AND ENERGY

THERMAL MANAGEMENT MATERIALS AND TECHNOLOGIES

Preface

Contributions from two Symposia and two Focused Sessions that were part of the 35th International Conference on Advanced Ceramics and Composites (ICACC), in Daytona Beach, FL, January 22–27, 2012 are presented in this volume. The broad range of topics is captured by the Symposia and Focused Session titles, which are listed as follows: Focused Session 1—Geopolymers and other Inorganic Polymers; Focused Session 2 - Computational Design, Modeling Simulation and Characterization of Ceramics and Composites; Focused Session 4—Advanced (Ceramic) Materials and Processing for Photonics and Energy. And Symposium 10—Thermal Management Materials and Technologies.

Focused Session 1 on Geopolymers and other Inorganic Polymers was the 10th continuous year that it was held. It continues to attract growing attention from international researchers and it is encouraging to see the variety of established and new applications being found for these novel and potentially useful materials. Six papers are included in this year's proceedings. The ability to decorate metakaolin with nanosilver particles is reported as the first step to biocidal porous geopolymer ceramics. The field of low pH, phosphoric acid reacting with fly ash to form chemically bonded phosphate ceramics is discussed. Such studies are welcome in the pursuit of sustainable and environmentally friendly ceramic composites.

Focused Session 2 was dedicated to design, modeling, simulation and characterization of ceramics and composites so as to further optimize their behavior and facilitate the design of new ceramics or composites with tailored properties. Thirty six technical papers were presented regarding the prediction of the crystal structure and properties of new ceramics, materials design for extreme/harsh environments, virtual materials design for new innovative materials, application of novel simulation methods for materials processing and performance, and the characterization and modeling of surfaces, interfaces, and grain boundaries at multiple scales. Six papers from this particular focused session are included in this issue.

The debut of Focused Session 4 was held during ICACC 2012. This session focused on synthesis, structural and functional characterization of self-organized materials and nanostructures of all ceramic materials with application potentials as functional materials, with particular consideration given to the capability to tailor and control material properties via surface and structural modifications. Two papers are included in this issue.

Symposium 10 discussed new materials and the associated technologies related to thermal management. This included innovations in ceramic or carbon based materials tailored for either high conductivity applications (e.g., graphite foams) or insulation (e.g., ceramic aerogels); heat transfer nanofluids; thermal energy storage devices; phase change materials; and a slew of technologies that are required for system integration. Two papers were submitted for inclusion in this proceedings issue.

The editors wish to thank the symposium organizers for their time and labor, the authors and presenters for their contributions; and the reviewers for their valuable comments and suggestions. In addition, acknowledgments are due to the officers of the Engineering Ceramics Division of The American Ceramic Society and the 2012 ICACC program chair, Dr. Sanjay Mathur, for their support. It is the hope that this volume becomes a useful resource for academic, governmental, and industrial efforts.

WALTRAUD M. KRIVEN, University of Illinois at Urbana-Champaign, USA
ANDREW GYEKENYESI, NASA Glenn Research Center, USA
GUNNAR WESTIN, Uppsala University, SWEDEN
JINGYANG WANG, Institute of Metal Research, Chinese Academy of Sciences, CHINA

Introduction

This issue of the Ceramic Engineering and Science Proceedings (CESP) is one of nine issues that has been published based on content presented during the 36th International Conference on Advanced Ceramics and Composites (ICACC), held January 22–27, 2012 in Daytona Beach, Florida. ICACC is the most prominent international meeting in the area of advanced structural, functional, and nanoscopic ceramics, composites, and other emerging ceramic materials and technologies. This prestigious conference has been organized by The American Ceramic Society's (ACerS) Engineering Ceramics Division (ECD) since 1977.

The 36th ICACC hosted more than 1,000 attendees from 38 countries and had over 780 presentations. The topics ranged from ceramic nanomaterials to structural reliability of ceramic components which demonstrated the linkage between materials science developments at the atomic level and macro level structural applications. Papers addressed material, model, and component development and investigated the interrelations between the processing, properties, and microstructure of ceramic materials.

The conference was organized into the following symposia and focused sessions:

Symposium 1	Mechanical Behavior and Performance of Ceramics and Composites
Symposium 2	Advanced Ceramic Coatings for Structural, Environmental, and Functional Applications
Symposium 3	9th International Symposium on Solid Oxide Fuel Cells (SOFC): Materials, Science, and Technology
Symposium 4	Armor Ceramics
Symposium 5	Next Generation Bioceramics

Symposium 6	International Symposium on Ceramics for Electric Energy Generation, Storage, and Distribution
Symposium 7	6th International Symposium on Nanostructured Materials and Nanocomposites: Development and Applications
Symposium 8	6th International Symposium on Advanced Processing & Manufacturing Technologies (APMT) for Structural & Multifunctional Materials and Systems
Symposium 9	Porous Ceramics: Novel Developments and Applications
Symposium 10	Thermal Management Materials and Technologies
Symposium 11	Nanomaterials for Sensing Applications: From Fundamentals to Device Integration
Symposium 12	Materials for Extreme Environments: Ultrahigh Temperature Ceramics (UHTCs) and Nanolaminated Ternary Carbides and Nitrides (MAX Phases)
Symposium 13	Advanced Ceramics and Composites for Nuclear Applications
Symposium 14	Advanced Materials and Technologies for Rechargeable Batteries
Focused Session 1	Geopolymers, Inorganic Polymers, Hybrid Organic-Inorganic Polymer Materials
Focused Session 2	Computational Design, Modeling, Simulation and Characterization of Ceramics and Composites
Focused Session 3	Next Generation Technologies for Innovative Surface Coatings
Focused Session 4	Advanced (Ceramic) Materials and Processing for Photonics and Energy
Special Session	European Union – USA Engineering Ceramics Summit
Special Session	Global Young Investigators Forum

The proceedings papers from this conference will appear in nine issues of the 2012 Ceramic Engineering & Science Proceedings (CESP); Volume 33, Issues 2-10, 2012 as listed below.

- Mechanical Properties and Performance of Engineering Ceramics and Composites VII, CESP Volume 33, Issue 2 (includes papers from Symposium 1)
- Advanced Ceramic Coatings and Materials for Extreme Environments II, CESP Volume 33, Issue 3 (includes papers from Symposia 2 and 12 and Focused Session 3)
- Advances in Solid Oxide Fuel Cells VIII, CESP Volume 33, Issue 4 (includes papers from Symposium 3)
- Advances in Ceramic Armor VIII, CESP Volume 33, Issue 5 (includes papers from Symposium 4)

- Advances in Bioceramics and Porous Ceramics V, CESP Volume 33, Issue 6 (includes papers from Symposia 5 and 9)
- Nanostructured Materials and Nanotechnology VI, CESP Volume 33, Issue 7 (includes papers from Symposium 7)
- Advanced Processing and Manufacturing Technologies for Structural and Multifunctional Materials VI, CESP Volume 33, Issue 8 (includes papers from Symposium 8)
- Ceramic Materials for Energy Applications II, CESP Volume 33, Issue 9 (includes papers from Symposia 6, 13, and 14)
- Developments in Strategic Materials and Computational Design III, CESP Volume 33, Issue 10 (includes papers from Symposium 10 and from Focused Sessions 1, 2, and 4)

The organization of the Daytona Beach meeting and the publication of these proceedings were possible thanks to the professional staff of ACerS and the tireless dedication of many ECD members. We would especially like to express our sincere thanks to the symposia organizers, session chairs, presenters and conference attendees, for their efforts and enthusiastic participation in the vibrant and cutting-edge conference.

ACerS and the ECD invite you to attend the 37th International Conference on Advanced Ceramics and Composites (http://www.ceramics.org/daytona2013) January 27 to February 1, 2013 in Daytona Beach, Florida.

MICHAEL HALBIG AND SANJAY MATHUR
Volume Editors
July 2012

Geopolymers and Other Inorganic Polymers

METAKAOLIN-NANOSILVER AS BIOCIDE AGENT IN GEOPOLYMER

José S. Moya[a], Belén Cabal[a], Jesús Sanz[a], Ramón Torrecillas[b]

[a]Instituto de Ciencia de Materiales de Madrid (ICMM-CSIC), Cantoblanco, 28049, Madrid, Spain
[b]Centro de Investigación en Nanomateriales y Nanotecnología (CINN), Consejo Superior de Investigaciones Científicas (CSIC) – Universidad de Oviedo (UO) – Principado de Asturias, Parque Tecnológico de Asturias, 33428, Llanera, Spain

ABSTRACT

Metakaolin is an aluminosilicate mineral product which is produced in quantities of several million tons per year worldwide and used in applications including supplementary cementitious materials in concretes, an intermediate phase in ceramic processing, as a paint extender, and as a geopolymer precursor. Meanwhile, geopolymers are also being considered for a variety of applications including low CO_2 producing cements, fiber-reinforced composites, refractories, and as precursors to ceramic formation.

Given the importance of metakaolin in several industry sectors, and also taking into account the possible large spectrum of applications, incorporating silver nanoparticles could provide an additional biocide functionality. In this regard, further studies of their structural evolution are required because of the presence of silver nanoparticles. For this purpose, this work is mainly focused on the evaluation of the effect of the presence of silver nanoparticles on kaolin/metakaolin structures and also on the study of their biocide capacity.

1. INTRODUCTION

Geopolymers are a class of inorganic polymers that are based on aluminosilicates. They are usually produced by adding a reactive aluminosilicate precursor, such as fly ash or metakaolin, to a highly alkaline silicate solution in order to facilitate the break-up of the primary aluminosilicate framework, leading to polymerisation and solidification. Then curing at 25-90 °C in a humid atmosphere completes the process. Geopolymers have received considerable attention because of their low cost, excellent mechanical and physical properties, low energy consumption and reduced "greenhouse emissions" at the elaboration process [1].

Metakaolin is preferred by the niche geopolymer product developers due to its high rate of dissolution in the reactant solution, easier control on the Si/Al ratio and the white colour. Metakaolin is formed by the dehydroxylation of kaolin. When kaolin is heated beyond the temperature of the dehydroxylation, endothermic metakaolin is formed. Between 500 and 900 °C, this is the main product obtained. The exact temperature range depends on the starting kaolin and on the heating regime.

Kaolin possesses a two-layered structure where a sheet of octahedrally coordinated aluminium is connected to a tetrahedrally coordinated silicon sheet. Sanz et al. [2] studied the kaolinite-mullite transformation by magic-angle spinning nuclear magnetic resonance (MAS-NMR) and determined the presence of Al in tetra- and pentacoordination in metakaolinite. The heat treatment at 700°C alters the structure of kaolin, the main change being the dehydroxylation of the octahedra. Above 800°C, tetrahedral sheets are broken making possible silica and alumina segregations. These modifications eliminate long-range order and make possible the formation at 980°C of amorphous mullite precursors. Dehydroxylation treatments cause the clay to become chemically reactive.

On the other hand, with the emergence and increase of microbial organisms resistant to multiple antibiotics, and the continuing emphasis on health-care costs, the development of materials with the ability to inhibit bacterial growth has been of great interest in recent years. The antimicrobial activity of silver has been known since ancient times. In the course of this work, a simple and fast method to prepare monodispersed silver nanoparticles embedded into kaolin and metakaolin is presented. These new silver-based nanostructured materials could have potentially wide-ranging applications, among others they could be used as precursors in geopolymer synthesis providing additional biocide functionality. Based on this, this work is mainly focused on the evaluation of the effect of the presence of silver nanoparticles on the structures of kaolinite/metakaolin and also on the study of their biocide capacity.

2. EXPERIMENTAL SECTION

2.1. Materials

Kaolin from CAVISA, La Coruña, Spain, with the following chemical analysis (wt.%): 54.3 SiO_2, 33.0 Al_2O_3, 0.19 TiO_2, 0.76 Fe_2O_3, 0.03 CaO, 0.37 MgO, 0.67 K_2O, 0.02 Na_2O, was used as raw material. Metakaolin was obtained after calcination of kaolin at 700 C for 24 h in air. Silver nanoparticles supported on kaolin were obtained using $AgNO_3$ as silver precursor and following two different reduction *vias*: the first by thermal reduction at 350 °C for 2 h in H_2 atmosphere and the second one by chemical reduction employing sodium borohydride as a reducing agent. In the case of metakaolin, only chemical reduction was performed.

2.2. TEM and FTIR characterization

The morphological aspects of the samples were studied by transmission electron microscopy (TEM) (Jeol microscope model FXII operating at 200 kV). Infrared spectroscopy was done in transmission in a vacuum atmosphere with a Fourier transform infrared spectrometer (Bruker IFS 66v/s).

2.3. MAS-NMR measurements

The ^{27}Al and ^{29}Si NMR spectra were obtained at room temperature, using an Avance (Bruker) spectrometer, operating at 104.3 MHz for ^{27}Al and 79.5 MHz for ^{29}Si (9.4 T external magnetic field). The samples were loaded in 4 mm rotors and spinned at 10 kHz during MAS-NMR spectra recording. In this study, $\pi/2$ (5 µs) pulses, 5s inter-accumulation intervals and 125 kHz filterings were used. All spectra were referred to TMS (tetramethylsilane) and $Al(H_2O)_6^{3+}$ as external standards. The error in chemical shift values was estimated to be lower than 0.5 ppm.

2.4. Antibacterial tests

Antibacterial tests were performed to investigate the effect of the kaolin/metakaolin/nAg powder on two different micro-organisms: *Escherichia coli JM 110* (Gram-negative bacteria), *Micrococcus luteus* (Gram-positive bacteria). The two different bacteria were incubated in a liquid media overnight at 37 °C. After that, 10 µL form each culture was diluted to 1 mL, using suitable media, and cultured at 37 °C for 6 h. The media used were Luria Bertani (LB). Subsequently, 150 µL of an aqueous suspension of kaolin/metakaolin/nAg composite (30 wt.%) was added to a final concentration of silver in each culture of 0.036 wt.%. A silver free media (a mixture of water containing the corresponding nutrient) was used as control. The microorganisms were tested for viability after culture on appropriate dilution from the corresponding cultures. The inocula were incubated at 37 °C with horizontal shaking for 48 h. The number of viable colonies was counted every 24 h.

3. RESULTS AND DISCUSSION

3.1. Characterization and effect of the presence of silver nanoparticles on the structure

The morphology of the samples was studied by TEM (Figure 1). A size distribution was carried out from different TEM images. There is a size distribution of globular-shaped silver nanoparticles that range between $d_{50} \sim 12 \pm 7$ nm in the case of kaolin samples and $d_{50} \sim 30 \pm 15$ nm for metakaolin. TEM images (Figure 1) also provide evidence for different distributions of silver nanoparticles, depending on the kind of support (i.e., kaolin or metakaolin) and on the reduction treatments used (for the different treatment see section 2.1). As it is clearly seen, in the case of the sample of chemically reduced kaolin (Figure 1.A), silver nanoparticles are anchored preferably on the crystal edges (the distribution of silver nanoparticles corresponds to 68 % at edges versus 32 % at basal surface), whereas in the case of the thermally reduced kaolin (Figure 1.B) and of the metakaolin (Figure 1.C), the distribution is more homogeneous (ca. 48 % at edges and 52 % at basal surface). Taking into account the structure of kaolin, silver nanoparticles could be bonded to the clays substrate *via* electrostatic interaction between the negatively charged and edge Al-O⁻ and Si-O⁻ groups of the surface clays, or with the anionic basal silicate planes if hydrogen bonds tightly linking contiguous layers are broken, before or during the incorporation of silver. It can be inferred that the adhesion of silver nanoparticles is closely related to the amount of surface hydroxyl groups located at the crystal edges. A decrease in hydroxyl groups of the clay leads to a more homogeneous distribution of silver nanoparticles, but with the disadvantage that their size is larger.

Although there are some studies in the literature about the synthesis of silver nanoparticles-kaolin composite materials [3, 4, 5], there are none on metakaolin particles. In order to obtain more information about the interaction of silver nanoparticles with metakaolin and whether or not its structure is modified, FTIR and MAS-NNR investigations have been performed.

The transformation of kaolin to metakaolin can be clearly deduced from the lattice region, 1400-400 cm⁻¹, of FTIR spectra (Figure 2). The kaolin starting material gives at least 10 well-defined IR bands in this region due to Si-O, Si-O-Al, and Al-OH vibrations: 1113, 1031, 1009, 699, 471, and 432 cm⁻¹ (Si-O); 938 and 912 cm⁻¹ (Al-OH); 792, 756, 539 cm⁻¹ (Si-O-Al$^{(VI)}$) [6]. The conversion to metakaolin totally removes these bands. In general, changes in the Si-O stretching bands and the disappearance of the Si-O-Al bands suggest strong modifications in tetrahedral and octahedral layers of the metakaolin. The incorporation of silver nanoparticles, both in kaolin and metakaolin, does not modify the positions of vibrational bands in the region 1100-400 cm⁻¹, indicating a small interaction between silicate layers and silver nanoparticles.

Magic-angle-spinning nuclear magnetic resonance (MAS-NMR) spectroscopy was employed with the aim of further clarifying the mode of bonding between the silver nanoparticles and the clay substrates. ^{29}Si-NMR is capable of distinguishing SiO_4 tetrahedra, with connectivity ranging from 0 to 4 (Qm species, with m standing for the number of bridging oxygens) [7]. In the original kaolin (Figure 3), two sharp absorptions at -91.5 ppm with full width at half maximum (FWHM) of 1 ppm were detected in ^{29}Si-NMR spectra, which corresponds to Si atoms in tetrahedral layers of layer silicates (Q^3 polymerization state) [2]. The incorporation of silver nanoparticles, following a chemical or thermal reduction (Figure 3), does not change the structure of kaolin. The spectra obtained in both cases are similar to the kaolin. Only in the case of the sample thermally reduced at 350 °C (Figure 3) it was observed a slight broadening of the main peak at -91.5 ppm and the formation of a new small broad band at -97 ppm. This could indicate the formation of a small fraction of amorphous silica. The ^{27}Al NMR spectrum of kaolin (Figure 4) shows a single peak at 0 ppm, characteristic of Al in octahedral coordination. No differences were detected when silver nanoparticles were supported on it chemically or thermally.

Metakaolin shows a very wide and asymmetric ^{29}Si resonance band (Figure 3) with two intense signals of different linewidth at -107.8 and at -101.4 ppm. This suggests the coexistence of aluminun free $Q^4(0Al)$ species, i.e. pure SiO_2 phase, and homogeneously distributed $Si(Al)O_4$ units $[Q^4(1Al)]$. When silver nanoparticles are supported on metakaolin (Figure 3), the intensity of the peak at -107.8 ppm is maintained but the peak at -101.4 ppm shifts to -103.6 ppm. From these results, a variation in the chemical environment of the silicon nuclei of the metakaolin upon incorporation of silver nanoparticles can be inferred.

The ^{27}Al NMR spectrum of metakaolin (Figure 4) contains tree peaks at 4, 28 and 54 ppm, attributed respectively to the presence of octahedral, pentahedral and tetrahedral aluminum [2]. The incorporation of silver nanoparticles to metakaolin stabilizes in some way the unstable structured of metakaolin. When spectra of starting and treated metakaolin are compared (Figure 4), a slight increase of hexacoordinated at expenses of tetrahedral and pentahedral aluminium is observed.

From the comparative analysis of ^{29}Si and ^{27}Al MAS-NMR spectra of starting materials and those of silver-kaolin or silver-metakaolin composites, it can be concluded that the silver particles interact preferentially with tetrahedral sheets and in a minor extent with octahedral layers of metakaolin particles. In this case some stabilization of hexacoordinated aluminium is favoured by incorporation of silver particles.

3.2. Biocide Activity

To investigate the antibacterial effect of n-Ag containing powders, a biocide test was performed innoculating 10^{10} colonies forming units into 1 mL of the corresponding medium with *Escherichia coli*, *Micrococcus luteus*. The microorganisms were tested for viability after culture on appropriate dilution from the corresponding cultures. During this test, the viable microorganisms were counted after 24 and 48 hours. As a control, silver free media (a mixture of water and the corresponding nutrient) was cultured under the same conditions.

The logarithm reduction (log η) has been used to characterize the effectivity of the biocidal agent:

$$\log \eta = \log A - \log B \tag{1}$$

where A is the average number of viable cells from innoculum controls after 24, 48 h, and B is the average number of viable cells from the substance after 24, 48 h.

As it can be seen in Figure 5 after 24 h, the presence of silver nanoparticles on kaolin surface, at the concentration of 0.036 wt.% of silver, reduces completely the number of colonies of *E.coli* and *M. luteus*, achieving a logarithm reduction higher than 10 which means a completely safe disinfection. Its high effectiveness is pointed out against both bacteria. In the case of metakaolin-silver nanocomposite, higher times are required to obtain quite similar results (i.e., log $\eta \sim 8$ for *E. coli* and log $\eta > 10$ for *M. luteus*). This behaviour could be attributed to the different sizes of silver nanoparticles in both materials. Previously, it was mentioned that silver nanoparticles embedded in kaolin are significantly smaller $[d_{50} \sim 12 \pm 7$ nm] than when they are in metakaolin $[d_{50} \sim 30 \pm 15$ nm]. The size of nanoparticle implies that it has a large surface area coming into contact with the bacterial cells and hence, it will have a higher percentage of interaction than the one with bigger particles.

The mechanism of bactericidal action of silver nanoparticles embedded into an inorganic matrix is still not fully known [8]. The silver nanoparticles show efficient antimicrobial property compared to other salts due to their extremely large surface area, which provides better contact with microorganisms. Sulphur-containing proteins in the membrane or inside the cells as well as phosphorous-containing elements are likely to be the preferential sites for silver nanoparticles binding. The nanoparticles release silver ions in the bacterial cells, which enhance their bactericidal activity [9].

The presence of low concentration of silver nanoparticles and silver ions involves a cytotoxic effect on human fibroblast stabling the cytotoxic limit above 30 ppm for silver nanoparticles and 1 ppm for silver ions [10]. In order to determine the toxicity of these materials, silver concentration in the resulting supernatant after 48 h was measured by Inductively Coupled Plasma (ICP) analyses. The results obtained are shown in Table 1. The concentration of silver nanoparticles in solution when metakaolin-nAg is used was about 4 times higher than in the case of kaolin-nAg, for all microorganisms studied. This can be due to the notable difference on particle size. Kaolin/metakaolin/nAg powders release significantly less fraction of silver keeping the concentration of silver below the cytotoxic level.

CONCLUSIONS

Following a simple and low-cost method it was possible to obtain silver dispersed nanoparticles attached to a kaolin/metakaolin matrix. The uneven distribution of the silver nanoparticles on the kaolin plates has been related to the presence of AlOH and SiOH at the edges of the plates. Conversely, silver nanoparticles are homogeneosly distributed in metakaolin. When ^{27}Al and ^{29}Si MAS-NMR spectra of two modified supports are compared, it was concluded that silver particles interact in a preferential way with tetrahedral layers of metakaolin materials. The bactericidal activity results show that these nanocomposites are strongly active against some of the most common Gram positive and negative bacterial strains, which could open the possibility to use them as precursors to fabricate biocide geopolymers with a large spectrum of applications.

ACKNOWLEDGEMENT

The authors acknowledge the financial support from ITMA, Spain (Ref. 010101090014) and the regional CAM Government (project S-2009/PPQ 1626).

REFERENCES

[1] P. Duxson, A. Fernández-Jiménez, J.L. Provis, G.C. Lukey, A. Palomo, J.S.J. van Deventer. Geopolymer technology: the current state of the art. Journal of Materials Science 42 (2007) 2917-2933.

[2] J. Sanz, A. Madani, J.M. Serratosa, J.S. Moya, S. Aza. Aluminun-27 and silicon-29 magic-angle spinning nuclear magnetic resonante study of the kaolinite-mullite transformation. Journal of American Ceramic Society 71 (1988) C-418-C421.

[3] K. Burridge, J. Johnston, T. Borrmann. Silver nanoparticles-clay composites. Journal of Materials Chemistry 21 (2011) 734-742.

[4] R. Patakfalvi, I. Dékány. Synthesis and intercalation of silver nanoparticles in kaolinite/DMSO complexes. Applied Clay Science 25 (2004) 149-159.

[5] R. Patakfalvi, A. Oszkó, I. Dékány. Synthesis and characterization of silver nanoparticles/kaolinite composites. Colloids and Surfaces A: Physicochemical and Enginneering Aspects 220 (2003) 45-50.

[6] E. Kristoff , A.Z. Juhasz, I. Vassanyi. The effect of mechanical treatment on the crystal structure and thermal behaviour of kaolinite. Clay and Clay Minerals 41 (1993) 608-612.

[7] G. Englehardt, D. Michael. High Resolution Solid-State NMR of silicates and zeolites. John Wiley & Sons Ltd. 1987.

[8] C. Marambio-Jones, E.M.V. Hoek, A review of the antibacterial effects of silver nanomaterials and potencial implications for human health and the environment, J. Nanopart. Res. 12 (2010) 1531-1551.

[9] M. Rai, A. Yadav, A. Gade, Silver nanoparticles as a new generation of antimicrobials, Biotechnol. Adv. 27 (2009) 76-83.

[10] A. Panáček, M. Kolář, R. Večeřová, R. Prucek, J. Soukupová, V. Kryštof, P. Hamal, R. Zbořil, , L. Kvítek, Antifungal activity of silver nanoparticles against Candida spp. Biomaterials. 30 [31] (2009) 6333-6340.

Correspondence address

Prof. José S. Moya
ICMM-CSIC
Sor Juana Inés de la Cruz, 3, 28049 Cantoblanco, Madrid, Spain
Tel. +34 91349000
Fax. +34 913720623
E-mail: jsmoya@icmm.csic.es

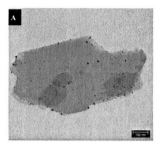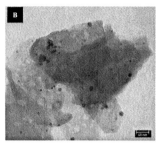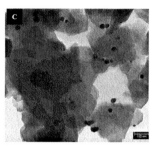

Figure 1. *TEM micrographs corresponding to: A) kaolin-nAg (chemically reduced), B) kaolin-nAg (thermally reduced) and C) metakaolin-nAg.*

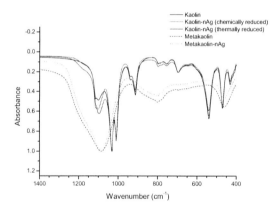

Figure 2. *FTIR spectra of kaolin´s and metakaolin´s sample.*

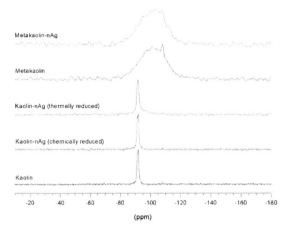

Figure 3. *Comparison of ^{29}Si MAS-ĮMR spectra of the samp les of kaolin and metakaolin.*

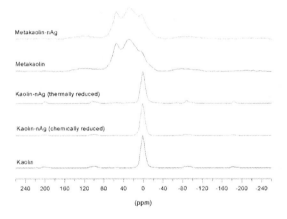

Figure 4. *Comparison of ^{27}Al MAS-ĮMR spectra of the samp les of kaolin and metakaolin.*

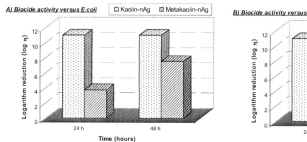
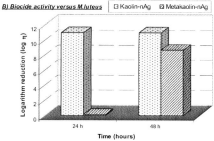

Figure 5. *Logarithm reduction* (log η) *of: A) E.coli and B) M.luteus at 24 and 48 h corresponding to kaolin-nAg and metakaolin-nAg.*

Table 1. *ICP analysis of silver lixiviated after 48 h of biocide test*

Bacteria strains	Ag lixiviated (ppm) Kaolin-nAg	Ag lixiviated (ppm) Metakaolin-nAg
E. coli (Gram -)	2.27	8.91
M. luteus (Gram +)	2.81	8.06

PARAMETERS THAT INFLUENCE SILICA DISSOLUTION IN ALKALINE MEDIA

A. Autef[a], E. Joussein[b], G. Gasgnier[c] and S. Rossignol[a]

[a] Groupe d'Etude des Matériaux Hétérogènes (GEMH-ENSCI) Ecole Nationale Supérieure de Céramique Industrielle, 12 rue Atlantis, 87068 Limoges Cedex, France.
[b] GRESE, EA 3040, 123 avenue Albert Thomas, 87060 Limoges, France.
[c] Imerys Ceramic Center, 8 rue de Soyouz, 87000 Limoges, France.
■ Corresponding author - sylvie.rossignol@unilim.fr – tel.: 33 5 87 50 25 64

ABSTRACT

Geopolymers have been the object of numerous studies because of their low environmental impact. The synthesis of these geomaterials is achieved by the alkaline activation of aluminosilicates. Alkaline activation is typically accomplished by the activation of potassium silicate or sodium silicate. These alkaline silicate solutions are relatively expensive. We thus attempted to create these solutions by the dissolution of potassium hydroxide and silica in water. This study focuses on the various parameters that can influence the dissolution of silica in basic media (pH>13). The samples used were amorphous silica, quartz sand and quartz ground to five different size distributions. The study of the dissolution of siliceous species was performed mainly by infrared spectroscopy by varying several factors. Stirring and solution volume played no significant role. However, the size distribution and crystallinity of silica were observed to significantly affect the kinetics of dissolution and the quantities of siliceous species in solution, which varied greatly according to the quantity of KOH introduced.

1. INTRODUCTION

Alkali silicates are currently the subject of significant interest because of their role as activators in the manufacture of geopolymers. Geopolymers are synthesized by the alkaline activation of alumino-silicates at ambient or higher temperatures [1]. The literature contains evidence that geopolymers based on potassium exhibit modified thermal and mechanical properties because of the larger size of the potassium ion compared to that of the sodium ion [2,3,4]. The use of silicate solution is very attractive because it plays an important role in the life cycle of these types of materials [5]. Two kinds of alkali silicate solutions may be used. The first is a ready-for-use commercial potassium silicate solution with a fixed Si / K molar ratio; however, this alternative presents an elevated cost. The second is a potassium silicate created by the dissolution of silica in KOH solution [6,7,8,9,10,11]. This dissolution allows the Si / K molar ratio of the potassium silicate to be modified. However, this method presents some inconveniences: the time required to complete the dissolution of the silica, the evaluation of the state of dissolution and the care required in handling the reagent (potassium hydroxide, volatile silica).

The formation of alkali silicate strongly depends on the rate of dissolution of the silica, which, in turn, depends on a variety of parameters, including the pressure and temperature [12], the silica size distribution [13], and the pH value [6,14]. Infrared spectroscopy is one of the most useful techniques for the characterization of alkaline solutions [15] because it verifies the presence of Si-O-Si bonds and provides information regarding the formation of OH bonds. Various silica compounds have been characterized by infrared spectroscopy in solid form [16] or in basic solution media [17]. Thus, the position of the Si-O, Si-O-M and

13

Si-O-Si vibration bands can be identified. These studies have elucidated the existence of several contributions within the vibration band of the Si-O-Si bond. These contributions are noted as Q^2, Q^3 and Q^4 and are situated at 1060, 1080 and 1165 cm^{-1}, respectively.

The purpose of this work was to investigate the influence of the silica size distribution, the influence of the solution volume and stirring and finally the influence of the Si / K ratio using infrared spectroscopy.

2. EXPERIMENTAL

Sample preparations

An alkali solution was used as the starting solution to dissolve silica. This solution was obtained by the dissolution of KOH pellets (85% purity) in water at room temperature. Various kinds of silica were used to allow a study of the dissolution: very fine amorphous silica (silica fume), quartz sand and quartz ground to five different size distributions. The characteristics of these silicate materials are reported in Table 1. The amorphous silica that was used was supplied by Sigma-Aldrich; it was very fine, and therefore very reactive. The various ground quartz samples supplied by the Imerys Ceramic Center were obtained by grinding the same raw material.

Table 1: Characteristics of the raw materials

Nature	Nomenclature	d_{50} (µm)	BET value (m^2/g)	Purity (%)
Amorphous silica	S0.1	0.14	202	99.8
Ground quartz	Q5	5.00	~ 2	98.6
Ground quartz	Q12	12.00	~ 2	98.9
Ground quartz	Q35	35.00	~ 1	98.8
Ground quartz	Q64	64.00	~ 1	98.9
Quartz sand	Q90	90.00	~ 1	98.0

To study the influence of experimental parameters (solution volume and stirring) on dissolution at room temperature, four mixtures (Si / K=0.7 and m / V=2.05) were prepared from the amorphous silica: a 35-mL alkali silicate solution with stirring (HV$_S$) and without stirring (HV) and a 10-mL alkali silicate solution with stirring (SV$_S$) and without stirring (SV). The influence of the silica size distribution on the obtained alkali silicate solutions (Si / K=0.7) was studied using various ground quartz samples dissolved at room temperature in the alkali solution by stirring. Finally, the amount of KOH used was modified to study the role of the Si / K molar ratio.

A commercial potassium silicate solution (Si / K=1.7, density 1.20, 76% water) was used as a standard. All of the solutions obtained were denoted by the nomenclature $Si_AK_t^Z$, where A represents the type of solid, t is the time of dissolution, and Z is the silica ratio. For example, the sample $Si_{S0.1}K_1^{0.7}$ is based on a potassium silicate solution obtained after a dissolution time of 1 h with S0.1 silica and a Si / K molar ratio of 0.7. The various nomenclatures are presented in Table 2.

Table 2: Nomenclature of various potassium silicate solutions

Nomenclature	Si/ K	Silica source	t (h)
SiKc	1.4	liquid	-
$Si_{S0.1}K_t^{0.7}$		S0.1	
$Si_{Q90}K_t^{0.7}$		Q90	
$Si_{Q5}K_t^{0.7}$	0.7	Q5	0-144
$Si_{Q12}K_t^{0.7}$		Q12	
$Si_{Q35}K_t^{0.7}$		Q35	
$Si_{Q64}K_t^{0.7}$		Q64	
$Si_{S0.1}K_{24}^{0.35}$	0.35		
$Si_{S0.1}K_{24}^{0.5}$	0.5		
$Si_{S0.1}K_{24}^{0.6}$	0.6		
$Si_{S0.1}K_{24}^{0.7}$	0.7	S0.1	24
$Si_{S0.1}K_{24}^{0.8}$	0.8		
$Si_{S0.1}K_{24}^{0.9}$	0.9		
$Si_{S0.1}K_{24}^{1}$	1.0		
$Si_{S0.1}K_{24}^{1.4}$	1.4		

Characterization

Fourier-transform infrared (FTIR) spectroscopy in ATR mode was used to monitor the silica dissolution. FTIR spectra were obtained using a Thermo Fisher Scientific 380 infrared spectrometer (Nicolet). The IR spectra were gathered over a wavenumber range of 400 to 4000 cm^{-1} with a resolution of 4 cm^{-1}. The atmospheric CO_2 contribution was removed with a straight line between 2400 and 2280 cm^{-1}. To allow comparisons of the various spectra, the spectra were corrected using a baseline and normalized. For each acquisition, a one-drop aliquot of solution was obtained; ten aliquots were obtained in total. The solution was stirred between any two acquisitions, except for the solutions HV and SV. Measurements were performed at different times at room temperature.

3. RESULTS

Characterization of raw materials

Silica raw materials (S0.1, Q5, Q12, Q35, Q64 and Q90) were first characterized by infrared spectroscopy in ATR mode (Fig. 1) to identify the different characteristic vibration bands initially present. The spectrum of ground quartz is presented (Q5) by itself because the spectra of Q12, Q35 and Q64 are identical; the differences between the spectra are due only to the size distribution. The IR bands and their assignments to different vibrations for each sample are reported in Table 3. For the Q90 and Q5 samples, the IR spectra show bands at similar positions, which indicates that the two samples are fundamentally similar in nature.

These bands correspond to the asymmetric stretching (v_{as}Si-O-Si) of the Si-O-Si (Q^4) bond at 1165 cm^{-1} [18], the asymmetric stretching (v_{as}Si-O-Si) of the Si-O-Si (Q^3) bond at 1080 cm^{-1} [19], the asymmetric stretching (v_{as}Si-O-Si) of the Si-O-Si (Q^2) bond at 1060 cm^{-1} [20], the bending of the Si-O bond at 805 cm^{-1} [21], the stretching of the quartz Si-O-Si bond at 777 cm^{-1}[22] and the stretching of the quartz at 693 cm^{-1}[19]. The same local order was observed for all of the ground quartz size distributions. For the S0.1 sample, only the asymmetric stretching (v_{as}Si-O-Si) of the Si-O-Si (Q^3) bond at 1080 cm^{-1} and the bending of the Si-O bond at 805 cm^{-1} were observed.

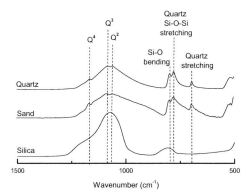

Figure 1: FTIR spectra of S0.1, Q90 and Q5 samples.

The spectra of the Q5 and Q90 samples consist of fine and well-defined peaks characteristic of crystallized phases; in contrast, the wide peaks observed in the spectrum of S0.1 are indicative of the sample's amorphous character. The observed contributions of the Si-O-Si bonds vary with the raw materials used. This result means that the contributions of Q^n, which would be related to distinct silicate species in solution, are different. Indeed, not only do siliceous materials present different size distributions, they also present different surface reactivities due to their respective amorphous or crystalline features. Thus, we focused our interest on silica, which displays an amorphous character and a fine size distribution (S0.1).

Table 3: Infrared bands of silica sources and their assignments [17]

Wavenumber (cm⁻¹)			Assignment of the IR band
Amorphous silica	Quartz sand	Ground quartz	
	693 w	694 s	Quartz: stretching
	777 m	777 s	Quartz: stretching
807 s, bd	796 m	796 s	Si-O bending
	1057 w	1059 s	v_{as}Si-O-Si (Q^2)
1073 w, bd	1083 w	1079 s	v_{as}Si-O-Si (Q^3)
1164 m, sh	1164 w	1163 m	v_{as}Si-O-Si (Q^4)

Measurement accuracy: ± 4 cm⁻¹; s, strong; m, medium; w, weak; bd, broad; sh, shoulder.

Kinetics of dissolution

IR spectroscopy measurements were performed at different times, from t=0 to 144 h after the addition of S0.1 to the KOH solution ($Si_{S0.1}K_t^{0.7}$). The various spectra recorded are presented in Fig. 2(A) along with the spectrum of SiKc (H_2O=76, 07%; SiO_2=16, 37%; K_2O=7, 56%). The OH and water bending vibration bands are observed at 3300 and 1650 cm⁻¹, respectively, and their intensity decreases with time in all of the spectra. The characteristic vibration bands of the Si-O-Si bond are located at approximately 1200-1000 cm⁻¹, and their intensities increase with time. Several contributions of the Si-O-Si bond (Q^2, Q^3 and Q^4) can be observed at 1060, 1080 and 1165 cm⁻¹, respectively [17]. The influence of silica's dissolution kinetics is also emphasized because the amount of siliceous species in solution increases with time. At the end, the final spectrum (t=144 h) was compared with that obtained using a commercial potassium silicate (SiK_C); the two spectra appear to be very similar.

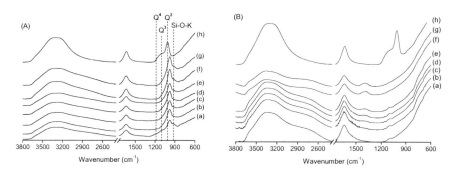

Figure 2: FTIR spectra of alkaline silicate solutions (A) $Si_{S0.1}K_t^{0.7}$ and (B) $Si_{Q90}K_t^{0.7}$ obtained with various silica substrates for different times (Si / K=0.7): (a) t=0, (b) t=0.25 h, (c) t=1 h, (d) t=5 h, (e) t=24 h, (f) t=50 h, (g) t=144 h, and (h) SiKc.

A similar study was performed in which S0.1 was replaced with Q90. The spectra recorded for $Si_{Q90}K_t^{0.7}$ (Fig. 2 (B)) do not show the same behavior as those obtained for S0.1. The vibration bands of the Si-O-Si bonds (1200-1000 cm^{-1}) do not evolve as a function of time, and the Si-OH band (1650 cm^{-1}) does not appear, which suggests that the Q90 solid has not been altered. After 24 h, the quartz sand had not dissolved, despite the solution having been stirred. Grains of quartz sand were observed in the solution. However, the kinetics of dissolution may be slower for Q90 than for S0.1 because of the large differences in size distribution and crystallinity. The reactive surface of Q90 in terms of its BET value (Table 1) is lower than that of S0.1; thus, it is more stable. Moreover, its stability is strengthened by its crystallinity, unlike silica, which is amorphous. Nevertheless, after 144 h of stirring, the solution ($Si_{Q90}K_{144}^{0.7}$) showed some changes: grains of quartz sand were always visible but the solution exhibited a viscosity that increased with time and quickly produced a silica gel. This behavior involved the presence of siliceous species in solution, but this hypothesis is not confirmed by the FTIR spectrum: the vibration bands characteristic of the silica still do not appear clearly, and only the presence of a shoulder is noted.

To understand the disappearance of bands when comparing the spectrum of $Si_{Q90}K_{144}^{0.7}$ and to that of SiKc in the 1500-to-500 cm^{-1} range, we focused on the Q^n contributions of the silicate species. The dissolution of the siliceous species was revealed by studying the evolution of the vibration modes (Q^2, Q^3 and Q^4 at 1060, 1080 and 1165 cm^{-1}, respectively) over time. The intensities of the peaks that correspond to the three vibration modes were noted, and the Q^2 / Q^3+Q^4 intensity ratio was then plotted according to time for the $Si_{Q90}K_t^{0.7}$ and $Si_{S0.1}K_t^{0.7}$ samples (Fig. 3). The ratio of the obtained values for Q90 are close to 1 and remain nearly constant over time, which suggests that the bonds at t=0 are the same as those at t=144 h. This hypothesis agrees with idea that no siliceous species dissolved with the quartz sand. However, in the presence of fine silica, the curve obtained for the $Si_{S0.1}K_t^{0.7}$ sample is shifted upward over time. The Q^2 vibration mode quickly dominates the Q^3 and Q^4 modes: a large number of siliceous species are present in solution after only a few minutes because of the strong reactivity of amorphous silica. The same calculation was performed for SiKc, and the value obtained was 0.818 (dotted line; Fig. 3). Finally, the percentage of water in SiKc is 76 %, whereas it is 53 % in the $Si_{S0.1}K_t^{0.7}$ and $Si_{Q90}K_t^{0.7}$ mixture. This high percentage of water in SiKc could be explained by the low value of the Q^2 / Q^3+Q^4 ratio due to the depolymerization of silicate species in the presence of excess water [17].

The remainder of this paper is concerned with the influence of external parameters such as stirring and solution volume.

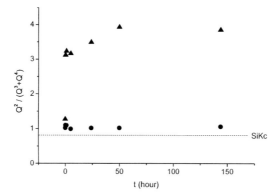

Figure 3. Intensity ratios $Q^2 / (Q^3+Q^4)$ determined from FTIR spectra of $Si_{S0.1}K_t^{0.7}$ ($\blacktriangle$), $Si_{Q90}K_t^{0.7}$ ($\bullet$) and SiKc (---) in contact with alkaline solution.

Influence of the experimental conditions: volume and stirring

FTIR experiments were performed to study the influence of the solution volume and stirring at different times (t=0-24 h) for a Si / K ratio of 0.7. Fig. 4(A) presents the FTIR spectra obtained after 0.25 h for each experiment. The spectra obtained for the four mixtures (HV, HV_S, SV and SV_S) show the same behavior; thus, it is difficult to assess the influence of the volume and stirring. The same feature is noted for other dissolution times; the corresponding spectra do not allow for the determination of the influence of stirring and the solution volume. Therefore, for each spectrum, the intensity of the OH band at 3300 cm^{-1} [15] and that of the Si-O-M vibration band located at 980 cm^{-1} [23] were noted to plot the intensity ratio (I_{Si-O-M} / I_{OH}) for each experiment with respect to time. The lack of variation with time in each case (Fig. 4(B)) indicates that the volume and stirring have no influence: the ratios remain constant in time and are almost equal in each case, i.e., we can infer that the chemical reaction is very fast. Indeed, the infrared spectra obtained at t=0.25 h are similar to those obtained for SiKc. These results suggest that, after a short time, the solution $Si_{S0.1}K_t^{0.7}$ contains the same silica species. Nevertheless, we have no information about the ionic silicate species. This study shows that the volume of the solution and stirring have no influence on the dissolution of the amorphous silica. However, the study of the kinetics of dissolution for S0.1 and Q90 reveals differences linked to the silica size distribution. Thus, in the following section, we focus on the influence of the silica size distribution on the dissolution of silica without variation in crystallinity.

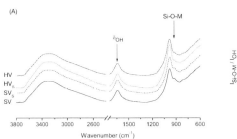
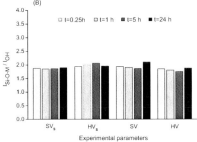

Figure 4. (A) FTIR spectra of HV, HV$_S$, SV$_S$ and SV at room temperature (t=0.25 h, Si / K=0.7); (B) evolution of Si-O-M / OH intensity ratio with the solution volume and stirring at different times.

Influence of the silica size distribution

The influence of the silica size distribution was studied by FTIR spectroscopy for ground quartz samples Q5, Q12, Q35 and Q64 at different times after the addition of silica (t=0 to 144 h). The Si / K molar ratio was 0.7. As before, the intensity ratios (I_{Si-O-M} / I_{OH}) were calculated and plotted according to the silica size distribution (Fig. 5).

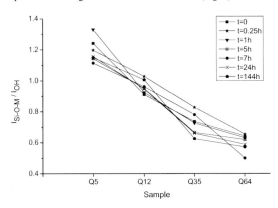

Figure 5. Evolution of Si-O-M / OH intensity ratio with silica size distribution for different times of reaction at room temperature (Si / K=0.7).

At t=0, the I_{Si-O-M} / I_{OH} ratio decreases with increasing silica size distribution. The same evolution is observed for all other times of dissolution. The I_{Si-O-M} / I_{OH} ratio is generally reduced by a factor of 1.85 when the size distribution increases from 5 to 64 μm. Wettability was measured to determine if a link existed between these reported intensities and the size distribution of the ground quartz. This experiment involved determining the volume of water required to saturate 1 g of dried material. The obtained results are presented in Table 4. The ratio of the wetting of the Q5 and Q64 mixtures is 1.91. This value is similar to that obtained using the I_{Si-O-M} / I_{OH} ratios and suggests that the quantity of species in solution is inversely proportional to the size distribution of the ground quartz. The decrease in the I_{Si-O-M} / I_{OH} ratios due to a decrease in the I_{Si-O-M} values with the size distribution could be explained by the fact that, when the silica size distribution increases, the reactivity of the surfaces is reduced, which produces a lower fraction of attacked silica. Therefore, few siliceous species pass into solution.

Therefore, the increase in the size distribution does not appear to prevent dissolution, but it plays an important role in the dissolution kinetics. Indeed, the kinetics of dissolution become slower as the size distribution of silica increases.

Table 4: Wetting values of various ground quartzes: Q5, Q12, Q35 and Q64

Silica	Size distribution (µm)	Wetting (µL/g)
Q5	5	480
Q12	12	360
Q35	35	280
Q64	64	250

Role of the Si / K molar ratio

The silica used in this study was amorphous silica, denoted as S0.1. To study the influence of the Si / K molar ratio on the formation of siliceous species, eight solutions were prepared (Si / K=0.35, 0.5, 0.6, 0.7, 0.8, 0.9, 1 and 1.4) and evaluated by FTIR at different times after the addition of the silica: t=0, 0.25, 1, 5 and 24 h. Fig. 6 presents the spectra obtained for the various Si / K molar ratios after 24 h of dissolution.

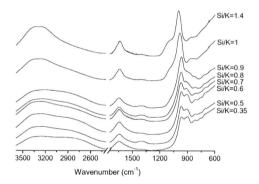

Figure 6. FTIR spectra of $Si_{S0.1}K_{24}{}^{Z}$ as a function of time for various Si / K ratios at t=24 hours.

For a Si / K = 1.4 ratio, the obtained spectrum shows a weak peak at 1080 cm^{-1} characteristic of the Si-O-Si bond observed in S0.1. The presence of this peak indicates that the silica is not dissolved because the quantity of KOH is very low. Only a small portion of S0.1 was dissolved. However, when the Si / K ratio is decreased and becomes closer to 0.35, the amount of KOH increases significantly. The intensity of the peak of the Si-O-Si bond observed previously decreases strongly, and a new peak appears at 980 cm^{-1}. The intensity of this second peak increases gradually when Si / K decreases. This trend corresponds to the vibration band of the Si-O-K bond. These two variations in intensity are due to the dissolution of the silica, which is, in turn, due to the increase in the amount of OH$^-$ species. The attack by KOH breaks the Si-O-Si bonds, and Si-O-K bonds form. Indeed, Bunker [24] has shown that tetrahedral sites common to all silicate glasses may undergo nucleophilic attack by OH$^-$ to first form a five-coordinate intermediate, which can then be broken down by the breaking of the Si-O-Si bonds (Equation 1).

$$\begin{matrix} & OH & & & OH & & & \\ & Si\text{-}O\text{-}Si\text{-}OH & + & OH^- & \longrightarrow & Si\text{-}O\text{-}Si\text{-}OH & \longrightarrow & Si\text{-}O^- + Si(OH)_4 & (1) \\ & OH & & & OH\ OH & & & \end{matrix}$$

According to equation 1, the presence of an increasing number of hydroxide ions in the attacker solution increases the formation of the intermediate, which leads to additional breaking of the Si-O-Si bonds.

To take advantage of all of the obtained spectra, the ($I_{Si\text{-}O\text{-}Si}$ / I_{OH}) intensity ratios were calculated and then plotted for each value of Si / K as a function of time (Fig. 7). It then became possible to observe the evolution of the amount of siliceous species in solution according to time and also according to the Si / K ratio. When the amount of KOH decreases (i.e., an increase in Si / K), the $I_{Si\text{-}O\text{-}Si}$ / I_{OH} intensity ratio increases with time. It would thus appear that, after some time, the same type of silica species (i.e., a decrease in the Si-O-Si peak intensity) and consequently the same reactivity would not be observed. Moreover, as the Si / K ratio increases, the $I_{Si\text{-}O\text{-}Si}$ / I_{OH} intensity ratio decreases over the same time interval. This result can be explained by the fact that, when the quantity of KOH decreases, fewer siliceous species are formed. In addition, fewer Si-OH bonds are present. Two regimes can be observed, depending on the Si / K ratio. When the Si / K ratio is between 1 and 1.4, silica is dissolved with the preferential formation of Q^3. In contrast, when this ratio is less than 0.5, the formation of Q^2 dominates. Therefore, to conclude, the type of siliceous species formed depends on the Si / K ratio. These results are in agreement with those obtained during the study of the dissolution of a glassy matrix in the presence of a basic solution to obtain a solution of water glass. The results demonstrate that dissolution is favored by an increase in the concentration of the alkaline solution.

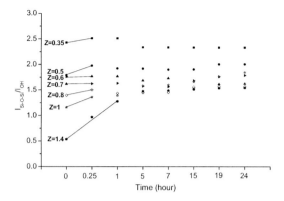

Figure 7. Intensity ratio of Si-O-Si / OH for $Si_{S0.1}K_{24}{}^Z$ as a function of time for various Si / K ratios at t=24 h.

4. CONCLUSION

The development of geopolymeric materials is responsible for an increasing consumption of alkali silicate solutions. Consequently, it is imperative to identify the means of controlling the alkaline solution and to understand all of the phenomena and reactions involved during geopolymer synthesis. In this study, several parameters that influence silica dissolution were determined. The size distribution of silica plays an important role with respect to the kinetics of dissolution. Indeed, for a given dissolution time, the rate of dissolution decreases as the size of the silica particles increases. The Si / K molar ratio is a secondary factor that affects the dissolution of the silica. When the Si / K ratio increases (i.e., the amount of KOH decreases), the partial dissolution of silica is approached due to a lack of KOH. When Si / K decreases, sufficient KOH is present to break the Si-O-Si bonds and completely dissolve the silica.

In contrast, the solution volume and stirring of the solution have no impact on the kinetics or the rate of dissolution. Identical alkali silicate solutions are obtained with or without stirring.

REFERENCES

[1] J.Davidovits, Chemistry and Applications, second ed. Geopolymer Institute, Saint Quentin, France, (2008).
[2] P. Duxson, J.L. Provis, G.C. Lukey, S.W. Mallicoat, W.M. Kriven, J.S.J. Van Deventer, Understanding the relationship between geopolymer composition, microstructure and mechanical properties, *Colloids and Surfaces A, 269, nos.1-3, 47-58*,(2005).
[3] B.G. Nair, Q. Zhao, R.F. Cooper, Geopolymer matrices with improved hydrothermal corrosion resistance for high temperature applications, *Journal of Material Science 42, 3083-3091*, (2007).
[4] J.W. Phair and J.S.J. Van Deventer, Effect of silicate activator pH on the leaching and material characteristics of waste-based inorganic polymers, *Minerals Engineering, 14, 3, 289–304*, (2001).

[5] B.C. McLellan, R.P. Williams, J. Lay, A. van Riessen, G.D. Corder, Costs and carbon emissions for geopolymer pastes in comparison to ordinary portland cement, *Journal of Cleaner Production, 19,1080-1090*, (2011).

[6] S.S Kouassi, Étude de la dissolution d'un réseau silicaté en présence d'une solution alcaline. *Thèse de doctorat , université de Limoges*, (2011).

[7] A. Paul, "Chemistry of Glasses". Ed. Chapman and Hall, New York, (1982).

[8] G.W MacLellan, E.B. Shand, Glass engineering handbook, Mc Graw-Hill, New york, (1984).

[9] J. Barton and C. Guillemet, LE VERRE, science et technologie, EDP sciences, (2005).

[10] B.M.J. Smets, On the mechanism of the corrosion of glass by water, *Philips Tech.Rev., 42, 59-64* (1985).

[11] Y. Niibori, M. Kunita, Dissolution rate of amorphous silica in high alkaline solution, *Journal of Juclear Science and Technology, 37, 349-357* , (2000).

[12] J.D. Hunt, A. Kavner, E.A. Schauble, D. Snyder, C.E. Manning, Polymerization of aqueous silica in H_2O-K_2O solutions at 25-200°C and 1 bar to 20kbar, *Chemical Geology, 283, 161-170*, (2011).

[13] P. Hrna, J. Marcial, Dissolution retardation of solid silica during glass-batch malting, *Journal of Jon-Crystalline Solids, 357, 2954-2959* , (2011).

[14] M. Dietzel, Dissolution of silicates and the stability of polysilicic acid, *Geochim. Cosmochim. Acta 64, 19, 3275*, (2000).

[15] P. Innocenzi, Infrared spectroscopy of sol-gel derived silica-based films: a spectra-microstructure overview, *Journal of Jon-Crystalline Solids, 316, 309-319* , (2003).

[16] R.J Bell, P. Dean, Atomic Vibrations in Vitreous Silica, *Discussion of the Faraday Society, 50, 55-61*, (1970).

[17] M. Tohoué Tognonvi, S. Rossignol, J.P. Bonnet, Effect of alkali cation on irreversible gel formation in basic medium, *Journal of Jon-Crystalline Solids, 357, 43-49* , (2011).

[18] W. K. W. Lee, J. S. J. Van Deventer, Use of Infrared Spectroscopy to Study Geopolymerization of Heterogeneous Amorphous Aluminosilicate, *Langmuir 19, 8726-8734*, (2003).

[19] M. Criado, A. Polomo, A. Fernandez-Jiménez, Alkali activation of fly ashes. Part 1 : Effect of curing conditions on the carbonation of the reaction products, *Fuel, 84, 2048-2054*, (2005).

[20] M.A. Muroya, Correlation between the formation of silica skeleton structure and Fourier transform reflection infrared absorption spectroscopy spectra, *Colloids and Surface A: Physicochemical and Engineering Aspects, 157, 147-155*, (1999).

[21] M. Hino, T. Sato, Infrared absorption spectra of silica gel-H216O, D216O, and H218O systems, *Bull. Chem. Soc. Jpn. 44, 33-37*, (1971).

[22] T. Fuss, A. Mogus-Milankovic, C.S. Ray, C.E. Lesher, R. Youngman, D.E. Day, Ex situ XRD,TEM,IR, Raman and NMR spectroscopy of crystallization of lithium disilicate glass at high pressure, *Journal of Jon-crystalline solids, 352, 4101-4111* , (2006).

[23] E. Prud'homme, P. Michaud, E. Joussein, J.-M. Classens, S. Rossignol, Role of alkaline cations and water content on geomaterial foams : Monitoring during formation, *Journal of Jon-crystalline solids, 357, 1270-1278* , (2011).

[24] B.C Bunker, D.R Tallant, T.J. Headley, G.L. Turner and R.J. Kirkpatrick, The structure of leached sodium borosilicate glass, *Physics and Chemistry of Glasses, 29, 106*, (1988).

HUMIDITY EFFECTS ON THE COMPLETION OF GEOPOLYMERIZATION IN DILUTE EVAPORATIVE SLURRIES

Brayden E. Glad and Waltraud M. Kriven
Materials Science and Engineering Department, University of Illinois at Urbana-Champaign, Urbana, IL, USA

ABSTRACT

Geopolymers provide an attractive alternative to existing materials for a variety of uses, but understanding of the geopolymerization reaction is still incomplete, particularly in real-life situations. Here, the effects of humidity or arid conditions on the completion of geopolymerization of both traditional sodium geopolymer formulations and dilute slurries are investigated. Characterization with Fourier transform infrared spectroscopy and X-ray diffraction quantitatively compare the extent of reaction, which is observed to be incomplete for dilute slurries, and that a portion of the alkali in these systems instead produces sodium carbonate monohydrate (thermonatrite). A preparatory step of 'pre-mixing' for several days the geopolymer reagents in a denser mixture followed by dilution is presented. This preparation is found to be useful in partially mitigating this effect. Possibilities for the kinetic causes for this result are briefly discussed.

INTRODUCTION

An alkali-activated condensate aluminosilicate or geopolymer (GP)[1-4], as defined by Joseph Davidovits, has been intensively investigated as a Portland cement alternative due to its high compressive strength[5] and ability to be synthesized from a variety of cheap precursors, including fly ash[6], slag[7], various clays, and even agricultural wastes[8]. When cured at ambient or near-ambient conditions, they demonstrate X-ray amorphous characteristics, and consist of AlO_4^- and SiO_4 tetrahedra balanced by group I cations[1]. Additionally, this class of materials demonstrates ceramic-like properties, including fire resistance, while requiring no high-temperature processing or sintering. However, the use of geopolymers is limited in many applications due to suffering failure in cracking under various drying conditions.

While geopolymer research has been conducted in humid and arid environments[3], little to no research has investigated the effects of humidity and water vapor conditions on geopolymerization. Most investigations have used sealed containers or molds in geopolymer preparation, resulting in substantial information on the geopolymerization process at effectively saturated conditions and little otherwise.

The effect of water content has been previously investigated under sealed conditions, identifying microstructural changes in porosity[9] and morphology[10]. However, for many practical applications, including bulk uses such as construction as well as coatings, a geopolymer formulation that does not require sealing might provide advantages over the traditional formulations. This investigation intended to determine what sort of changes in processing were required to account for this difference, since it was observed that drying geopolymers in arid conditions leads to failure by cracking[4]. Further, it is not uncommon for anomalous compressive or flexure strengths to be reported in the literature, particularly as its primary focus has been in optimization. It is possible that changes in ambient humidity conditions might be responsible for some strength differences reported for identical formulations.

Opening the material to the environment allows for slurry viscosity reduction through increased initial water content. However, the observed qualitative differences between sealed samples and samples cured under low-humidity conditions poses a question as to what precisely is the effect of water loss during curing. While obviously geopolymerization can be inhibited through dilution, the question of whether evaporation of excessively dilute geopolymer slurry down to normal synthesis conditions would result in the same or different microstructure as a regularly cured geopolymer had not been investigated.

EXPERIMENTAL
Synthesis

Geopolymers were synthesized from as-received metakaolin (Metamax EF, BASF Corp.) and sodium silicate solution. Sodium silicate solution (waterglass) was prepared as follows: 160.0 g sodium hydroxide (Sigma-Aldrich) was added to 360.0 g DI water in a stainless steel container and stirred using a polytetrafluoroethane stir bar with a magnetic stir plate until fully dissolved. The container was then sealed with plastic film, and the temperature of the stir plate was set to approximately 45 °C. A total of 240.0 g Fumed silica (Cab-o-sil, Cabot Corp.) was then added in small batches of approximately 10 g, each time allowing for complete or almost complete dissolution of the silica before adding the next batch, with the container remaining sealed when not adding silica. After all silica was added, the container was resealed and the contents were allowed to stir at the elevated temperature ($\approx$45 °C) for 24 hours. After 24 hours, the solution of sodium silicate was poured into a polyethylene container, and allowed to equilibrate sealed at ambient conditions for at least one month.

Three types of samples were synthesized, which are being labeled as dilute evaporative geopolymers, premixed geopolymers and controls. For the dilute evaporative geopolymers, a mass of 2.78 g metakaolin (Metamax EF, BASF Corp.) was placed in a polypropylene container (Solo Cup Co.) and 40.0 g of DI water was added. Then 4.75 g waterglass was added to the mixture, producing an initial nominal $Na_2O \cdot Al_2O_3 \cdot 4(SiO_2) \cdot 197(H_2O)$ formulation (precise ratio 1.06·1·4.22·197.4 when impurities in the metakaolin are considered). The sample was ultrasonicated (W385 Sonicator, Heat Systems) at 240 watts while stirred on a stir plate until well-mixed (5 minutes). The sample was then placed in a controlled humidity chamber (TestEquity 1000H) or in a sealed container with a large excess of calcium sulfate (Drierite, W. A. Hammond Co.), and maintained at 25 °C until mass loss ceased.

For the premixed geopolymers, 11.1 g metakaolin, 0.51g $\pm$ 0.02 g Darvan 821A dispersant (40% ammonium polyacrylate in water, R. T. Vanderbilt Co.), and 19.0 g waterglass were added simultaneously to a variable quantity of DI water (5.55 g, 9.71 g, 19.43 g, and 22.20 g), and sealed, then stirred using a polytetrafluoroethane stir bar with a magnetic stir plate for a specified period (1, 3, and 7 days). After that period, a quantity of material was extracted to duplicate the mass of the dilute evaporative geopolymers, and then diluted to the same nominal $Na_2O \cdot Al_2O_3 \cdot 4(SiO_2) \cdot 197(H_2O)$ formulation as above with DI water, receiving the same sonication, humidity and temperature conditions until mass loss ceased. Previous experiments identified no effect of the presence or absence of dispersant on the dilute evaporative geopolymers, but the dispersant was useful in managing the viscosity of the premixes prior to dilution.

Control geopolymers were synthesized in accordance with previous work[4] with 5.55 g metakaolin (Metamax HRM, BASF Corp.) and 9.25 g waterglass, producing a nominal $Na_2O \cdot Al_2O_3 \cdot 4(SiO_2) \cdot 11(H_2O)$ formulation (precise ratio 1.03·1·4.17·11.28 when impurities in the metakaolin are considered) and mixed in a planetary centrifugal mixer (ARE-250, Thinky Corp.) at 1000 rpm for 120 seconds, with a 60 second debubbling at 1200 rpm. The dense slurry was then placed in a polypropylene container under humidity conditions described above.

Procedure

This experiment was conducted at 25 °C and desiccant (excess anhydrous calcium sulfate, Drierite Co.) conditions to both closely simulate geopolymer curing under arid conditions and more clearly identify differences. The direct comparison of standard sealed and arid-cured geopolymers as a function of initial water content was straightforward. Samples were allowed to cure under these conditions, with desiccant changed as necessary, until mass loss was no longer observed.

Alkali dependence was investigated using the same dilute slurry formulation, with additional sodium hydroxide immediately added to reach the standard $Na_2O \cdot 11H_2O$ proportion. This resulted in a nominal composition of approximately $18(Na_2O) \cdot Al_2O_3 \cdot 4(SiO_2) \cdot 197(H_2O)$.

The premixed samples were taken from each pre-mixture after 1, 3 and 7 days, and then diluted, processed and allowed to cure as described above. The desiccant was maintained such that its rated capacity exceeded 150% of the water content of the samples, and an excess of calcium oxide was also placed in the drying chamber to mitigate the effects of carbon dioxide during curing. The drying chamber was kept closed except for measuring, but was incompletely airtight.

Characterization

Cu Kα X-ray powder diffraction (XRD) was conducted with a Seimens-Bruker D5000 diffractometer. Tests were conducted at 40 kV, 30 mA, a step size of 0.02° 2-Θ, and a measurement time of 8 seconds/step. Diffuse reflection infrared spectra were taken using a Nicolet Nexus 670 spectrometer on powdered samples mixed with KBr (1:19 wt ratio). Powder samples for both instruments were generated by dry grinding in an alumina mortar and pestle, and were collected after being passed through a 44 μm mesh.

RESULTS

It was found that the highly alkali $18(Na_2O) \cdot Al_2O_3 \cdot 4(SiO_2) \cdot 197(H_2O)$ dilute slurries did not fully evaporate or form a coherent solid. In every other case, monolithic samples eventually formed, but the mechanical properties, microstructure and phase composition of the monoliths varied considerably. All of the monoliths cured under the desiccant conditions experienced cracking failure, as expected. However, because of the massive excess of water used in these slurries, it is unsurprising that maximizing the humidity for even two full days (95% humidity) prior to reducing the humidity to desiccant conditions had no meaningful effect on the results, but increasing the humidity to 70% and 30% for the duration of curing resulted in a solid monolith with minimal cracking.

As seen in the x-ray diffraction data in Figure 1, the characteristic shift of the large geopolymer amorphous hump from ≈22° 2-Θ to ≈28° 2-Θ was not observed in the highly dilute geopolymer slurries. Instead, there is a large amorphous hump at ≈25° 2-Θ and unreacted metakaolin visible at ≈22° 2-Θ. Additionally, the presence of sodium carbonate monohydrate (thermonatrite, ICDD No. 04-009-3774) indicates that this phase was formed in parallel to the geopolymer reaction. In every case, the presence of carbonate was associated with an incomplete reaction. The sharp peak at ≈25.2 2-Θ is indicative of anatase (ICDD No. 00-21-1272), which appeared in every micrograph and is the result of titania impurities in the metakaolin.

Infrared spectroscopy data reflect the same result, as seen in Figure 2. In the controls, there is a clearly visible characteristic peak shift of the ≈1080 cm^{-1} Si-O-Si symmetric stretch to a ≈1000 cm^{-1} asymmetric stretch representing the interaction of the Si-O-Si bonds and the Al-O-Si stretch observed at ≈810 cm^{-1} [11]. In the samples produced from highly dilute slurries, such a shift was greatly reduced, and the carbonate peaks at ≈1460 cm^{-1} and ≈2900 cm^{-1} were substantially more prominent.

The premixed samples illustrated incomplete reactions as well, but demonstrated a pronounced difference in reactivity as measured by FTIR peak shift, as shown in Table 1. However, these samples produced similar diffraction profiles as the dilute slurries, with the amorphous hump position apparently unchanged. A limited amount of sodium carbonate decahydrate (natron, ICDD No. 00-037-0451) was also observed in these diffractograms, but only in samples premixed for more than one day.

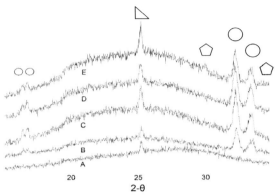

Figure 1: X-ray diffraction data of a geopolymer control (1·1·4·11 composition) (A), a dilute slurry without premixing (B), and the results of 18.1 mol water-1 day (C), 22.9 mol water-1 day (D) and 37.4 mol-7 day (E) premixes. Present in all the samples was anatase (00-21-1272, triangle). In all the dilute slurries thermonatrite (04-009-3774, circles) was observed and the amorphous hump at 28° 2-Θ is shifted to 25°. In the 7 day premixed samples traces of natron (pentagons) are present.

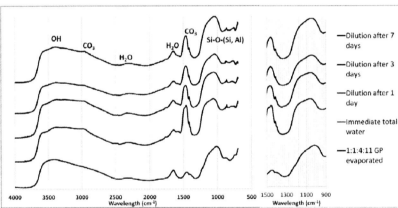

Figure 2: FTIR spectra of premixed evaporative geopolymers are compared. Dilute evaporative samples (197 mol water/mol GP) made with premixtures (34.2 mol water/mol GP) and without are compared to a standard dense geopolymer control. As seen in the closeup (right), premixing for an extended period shifts the Si-O-(Si, Al) stretch peak towards the right, indicating greater aluminate incorporation.

Table I: Position of primary Si-O-(Si, Al) stretch peak, as determined by FTIR. Values close to 1080 cm-1 indicate that the symmetric Si-O-Si stretch dominates. Premixes marked 'solid' solidified despite the mixing process and were thus unable to be used.

Sample Initial Water Content 1:1:4:x (mol/mol GP)	Si-O-M FTIR peak (cm^{-1})			
	Never diluted	Diluted to 1:1:4:197 after 1 day	Diluted to 1:1:4:197 after 3 days	Diluted to 1:1:4:197 after 7 days
11 (control, sealed)	1008	Solid	Solid	Solid
11 (control, evaporative)	1019	Solid	Solid	Solid
197.4 (dilute slurry)	1061	--	--	--
18.1	--	1056	Solid	Solid
22.9	--	1051	Solid	Solid
34.2	--	1058	1042	1044
37.4	--	1059	1057	1046

DISCUSSION

The fact that premixing appears to have improved reactivity suggests that some initial reacted material is preserved through the subsequent dilution and evaporation, even when the final solid-water ratios are similar. This could be useful for certain applications where viscosity must be minimized (such as spray coatings). The fact that premixing time had a greater influence than concentration over the ranges investigated in this experiment suggests that the range of premix concentrations could be expanded for further viscosity control.

The ability to form carbonate rather than geopolymer is also apparently a consequence of the differing concentrations of reagents in solution. Kinetic analysis of geopolymer[12] and similar systems[13] identify hydroxide concentrations as critical to dissolution of the aluminate species. The subsequent condensation reaction depends completely and approximately quadratically on the concentration of aluminosilicate oligomers in solution[12] while the formation of the competing carbonate salt precipitates should depend only linearly on carbonate presence.

Alternatively, sample temperature differences might significantly affect precipitation nucleation. Although all samples were present in 25 °C conditions, relatively rapid geopolymerization occurring in regular geopolymer controls increases the sample temperature significantly. At temperatures around 25 °C, thermonatrite precipitation occurs readily with minimal nucleation time[14], while at higher temperatures significant nucleation time is observed.

Creation of a dilute evaporative geopolymer with full reactivity proved difficult. Substituting the sodium aluminate for the less reactive metakaolin proved ineffective, as bayerite was produced in preference to geopolymer. For controls, Metamax EF solidified within the planetary mixer for any reasonable degree of mixing, so Metamax HRM, with larger particle size, was used. Likewise, the dispersant was omitted due to its phase separation in the planetary mixer. Other experiments indicated that there is no meaningful effect of these changes on geopolymerization.

CONCLUSION

Geopolymerization in unmodified metakaolin-sodium silicate slurries is strongly dependent on curing conditions, with arid conditions during curing producing a differing morphology than samples produced in a traditional sealed mold. The use of dilute slurry and an evaporative method to ensure adequate water under arid conditions resulted in the inhibition of geopolymerization, with thermonatrite being preferentially produced. This effect was successfully mitigated through extensively mixing more concentrated slurry, then diluting it before use, but even then, only partial geopolymerization was observed. Kinetic analysis can provide some insight into this observation, but analysis is complicated by the exothermic nature of the geopolymerization reaction itself.

ACKNOWLEDGEMENTS

This investigation was supported with funds from Dow Chemical Co., Midland, MI. This research was carried out in part in the Frederick Seitz Materials Research Laboratory Central Facilities, University of Illinois.

REFERENCES

[1] J. Davidovits, Geopolymers - Inorganic Polymeric New Materials, *J. Therm. Anal.*, **37**, 1633–56 (1991).

[2] J. Davidovits, Mineral Polymers and Methods of Making Them, US Patent 4349386 (1982).

[3] J. Davidovits, *Geopolymer Chemistry and Applications*, Institut Geopolymere: St. Quinten, Fr. (2008).

[4] W. M. Kriven, J. L. Bell and M. Gordon, Microstructure and Microchemistry of Fully-Reacted Geopolymers and Geopolymer Matrix Composites, in *Ceram. Trans.*, **153**, *Advances in Ceramic-Matrix Composites IX* (eds N. P. Bansal, J. P. Singh, W. M. Kriven and H. Schneider), American Ceramics Society, Westerville, OH (2003).

[5] P. Duxson, S. W. Mallicoat, G. C. Lukey, W. M. Kriven and J. S. J. van Deventer, The Effect of Alkali and Si/Al Ratio on the Development of Mechanical Properties of Metakaolin-Based Geopolymers, *Colloids Surface. A*, **292**, 8–20 (2007).

[6] J.G.S. van Jaarsveld, J.S.J. van Deventer and G.C. Lukey, The Effect of Composition and Temperature on the Properties of Fly Ash- and Kaolinite-based Geopolymers, *Chem. Eng. J.*, **89**, 63-73 (2002).

[7] S. A. Bernal, J. L. Provis, V. Rose and R. M. de Gutierrez, Evolution of Binder Structure in Sodium Silicate-Activated Slag-Metakaolin Blends, *Cement Concrete Comp.*, **33**, 46-54 (2011).

[8] S. Detphan and P. Chindaprasirta, Preparation of Fly Ash and Rice Husk Ash Geopolymer, *Int. J. Miner. Met. Mater.*, **16**, 720-6 (2009).

[9] T. L. Metroke, M. V. Henley and M. I. Hammons, Effect of Curing Conditions on the Porosity Characteristics of Metakaolin–Fly Ash Geopolymers, in *Strategic Materials and Computational Design: Ceramic Engineering and Science Proceedings*, **31** (eds W. M. Kriven, Y. Zhou, M. Radovic, S. Mathur and T. Ohji), John Wiley & Sons, Inc., Hoboken, NJ (2010).

[10] K. Okada, A. Ooyama, T. Isobe, Y. Kameshima, A. Nakajima and K. J.D. MacKenzie, Water Retention Properties of Porous Geopolymers for Use in Cooling Applications, *J. Eur. Ceram. Soc.*, **29**, 1917-23, (2009).

[11] V. F.F. Barbosa, K. J.D. MacKenzie and C. Thaumaturgo, Synthesis and Characterisation of Materials Based on Inorganic Polymers of Alumina and Silica: Sodium Polysialate Polymers, *Int. J Inorg. Mater.*, **2**, 309-17 (2000).

[12] J. Provis and J. van Deventer, Geopolymerisation Kinetics. 2. Reaction Kinetic Modeling, *Chem. Eng. Sci.*, **62**, 2318-29 (2007).

[13] A. Blum and A. Lasaga, Role of Surface Speciation in Low-Temperature Dissolution of Minerals, *lature* , **331**, 431-3 (1988).

[14] A. A. Shaikh, A. D. Salman, S. McNamara, G. Littlewood, F. Ramsay and M. J. Hounslow, In Situ Observation of the Conversion of Sodium Carbonate to Sodium Carbonate Monohydrate in Aqueous Suspension, *Ind. Eng. Chem. Res.*, **44**, 9921-30 (2005).

THE EFFECT OF BASALT CHOPPED FIBER REINFORCEMENT ON THE MECHANICAL PROPERTIES OF POTASSIUM BASED GEOPOLYMER

Sean S. Musil, Greg Kutyla and W. M. Kriven

Department of Material Science and Engineering, University of Illinois at Urbana-Champaign, Urbana, IL, 61801, USA

ABSTRACT

The processing and mechanical properties of potassium-based geopolymer ($K_2O \cdot Al_2O_3 \cdot 4SiO_2 \cdot 11H_2O$) composites containing 1/2 inch long basalt chopped fiber reinforcement have been evaluated and compared to previous data for 1/4 inch long basalt fiber reinforcement. Geopolymer composite test samples were hand fabricated at ambient temperature and cured in a constant humidity, 50°C temperature/humidity chamber. The effects of varying weight percent of fiber reinforcement and heat treatment temperature for fixed fiber content were examined in this study. Reinforcement by ten weight percent basalt chopped fibers significantly improved the room temperature average bend strength over that of pure geopolymer from roughly 2 MPa to 27 MPa. The composite was able to preserve 75% bend strength after heat treatment to 200°C and 12% strength after heat treatment to 600°C, with only five weight percent of fiber reinforcement. Addition of 1/2 inch long fibers showed 20-50% strength increases over the use of 1/4 inch fibers. A Weibull statistical analysis was performed showing a significant increase in reliability with increasing amounts of chopped basalt fiber reinforcement.

INTRODUCTION

Geopolymers are a type of inorganic polymeric material composed of alumina, silica, and alkali metal oxides. They have the advantage of being synthesized as a liquid, allowing them to be cast into any desired shape. They have a faster setting time compared to ordinary Portland cement, and have a smaller carbon footprint associated with production. They also compare favorably in strength.[1] Additionally, the abundance of aluminosilicate materials gives geopolymers the potential to be more cost effective than Portland cement.

One area that geopolymers have been found to excel is as a matrix for composites due to the ease of synthesis and reinforcement adhesion. Since geopolymers begin as liquids, the addition of reinforcement phases such as particles or fibers only requires mixing them with the geopolymer binder. Geopolymers have also shown promise as adhesives, bonding strongly to a wide range of ceramics, metals, and even polymers.[2,3] This property guarantees maximal load transfer from the geopolymer matrix to any reinforcement phase. With properly chosen fiber reinforcement, geopolymer strength can be doubled with as little as one weight percent fiber.[4]

Another useful property of geopolymers is their resistance to heat and oxidizing environments. In particular, potassium-based geopolymer remains chemically stable until 1000°C.[5] Between 1000 and 1100°C, the potassium geopolymer crystalizes into cubic leucite.[6] Leucite, a ceramic of composition $K_2O \cdot Al_2O_3 \cdot 4SiO_2$, is desirable in its own right as it retains strength up to 1200°C.[5] However; the geopolymer must first shed its water, doing so up to 400°C. This dehydration process proves quite destructive for unreinforced potassium geopolymer. Heated samples develop microcracks that increase in width and depth with increasing temperature. Certain compositions of unreinforced potassium geopolymer even turn to powder as temperature approaches 1000°C. Potassium geopolymer of other chemical compositions, for example potassium polysialate, have shown thermal stability up to 1400°C.[11] Another solution to this was found to be the addition of a reinforcement phase to hold the geopolymer together during dehydration. Additionally, a weak fiber-matrix interface may provide channels for more graceful dehydration.

Previous research by Lowry et al.[9] has been performed on sodium geopolymer with polypropylene (PP) fiber reinforcement. Two inch long PP fibers at 2.48 weight percent resulted in 18.3±0.8 MPa average flexure strength. While PP fibers are suitable for low temperature reinforcement, they cannot withstand high temperature applications. Additionally, research performed by Rill et al.[4] has also been performed on potassium geopolymer with 1/4 inch long basalt fiber reinforcement and has shown that the addition of basalt chopped fiber increases the flexure strength of the geopolymer and is suitable for higher temperature applications.

For this study, 1/2 inch long basalt chopped fiber was used as the reinforcement phase. The chemical composition of the potassium geopolymer matrix used in this research was identical to that used in the Rill work. Basalt is a blend of silicates with components generally being the oxides of aluminum, calcium, iron, magnesium, and sodium. The basalt fiber, Type BCS 13 1/3 KV02, manufactured by Kamenny Vek Ltd, Dubna, Moscow Region, Russia, has a diameter of 13 μm, a bulk density of 2.67 g/cm^3, a tensile strength of 2700-3200 MPa, a tensile modulus of 85-95 GPa and a softening temperature of 1050°C. Fibers had a composition of 55.7% silica, 15.4% alumina, 10.4% ferric oxide, 7.4% calcia, 4.1% magnesia, 2.4% soda, 1.5% potassia, and 1.2% titania, by mass.[10] The fibers were coated with a silane sizing, which was shown by Rill et al.[4] to produce composites with improved room temperature flexure strength as compared to those using fibers without sizing.

EXPERIMENTAL PROCEDURES

Potassium geopolymer binder was developed by mixing a potassium hydroxide, deionized water and fumed silica solution (potassium water-glass) with metakaolin clay using an Eirich mixer, Type EL1, Maschinenfabrik Gustav Eirich GmbH & Co KG, Hardheim, Germany. The mixer was run at 1200 rpm for 15 minutes to produce a homogeneous slurry. The Eirich mixer also allowed for the mixing bowl to rotate at 170 rpm (high setting) to aid in distributing the geopolymer constituents. The resulting geopolymer chemical composition was $K_2O \cdot Al_2O_3 \cdot 4SiO_2 \cdot 11H_2O$.

The geopolymer binder was vibrated to remove air bubbles introduce during mixing and then mixed with varying weight percent of 1/2 inch long chopped basalt fiber reinforcement. For this mixing step, the mixer was run at 300 rpm and the bowl at 85 rpm (low setting). These were the minimum allowable mixing speeds for the Eirich mixer. Every three minutes, the mixer was stopped to remove large clumps of fibers that had conglomerated around the stirring rod. This technique allowed the fibers tows to be separated and evenly distributed within the geopolymer binder.

The resulting composite mixture was then hand poured into a mold consisting of six, one inch by one inch by six inch bars. The mold was sealed tightly and wrapped with plastic wrap to avoid dehydration during curing. The samples were cured at 50°C and 70% humidity in a Test Equity 1000H Temperature/Humidity Chamber for 24 hours before being removed from the mold.

A total of 42 geopolymer composite samples were created from 7 batches; one batch at one weight percent, four batches at two weight percent and two batches at five weight percent chopped fiber reinforcement. Two and five weight percent batches were subjected to heat treatments to determine the residual strength of the composite after heating. The heat treatments were done in a Carbolite CWF 1200 box furnace with a Eurotherm 3216 temperature controller. The samples were heated and cooled at a ramp rate of ±5 °C/min, with no isothermal soak at the target temperature.

All of the samples were then subjected to three point flexural tests using an Instron Universal Testing Frame, following ASTM standard C78/C78M-10. The span length of the lower rigid supports was calculated as three times the average specimen thickness and the supports were placed equidistant from the point of load application. Loading rate was determined to be 216.75 N/min using ASTM C78/C78M-10 guidelines to maintain a 1.0 MPa/min target rate of stress increase on the tension face of the specimen.

RESULTS AND DISCUSSION

Flexure stress and strain data were collected and are summarized in Table I and Table II below. Average maximum flexure stress (strength) values were calculated per batch, as well as a standard deviation for that batch. The standard deviation, S, was calculated using equation (1) below; where x is the maximum strength of the sample, $\bar{x}$ is the average of the maximum strength values for that batch and N is the batch size of 6. Representative stress - strain curves for the room temperatures samples are shown in Figure 1.

$$S = \sqrt{\frac{\Sigma(x-\bar{x})^2}{N}}$$

(1)

Table I. Summary of three point flexure test results for room temperature samples.

Fiber Weight (%)	Max Strength (MPa)	Average Strength (MPa)	Standard Deviation (MPa)	Fiber Weight (%)	Max Strength (MPa)	Average Strength (MPa)	Standard Deviation (MPa)
0	3.207			5	19.477		
0	2.680	2.155	0.936	5	18.382		
0	1.309			5	14.044	16.758	2.215
0	1.425			5	15.272		
1	3.309			5	16.615		
1	3.140			7	23.042		
1	5.487	3.925	0.852	7	29.682		
1	4.209			7	21.261	24.243	3.912
1	3.797			7	27.260		
1	3.609			7	19.076		
2	4.478			7	25.137		
2	7.199			10	25.337		
2	8.732	7.256	1.596	10	24.937		
2	8.772			10	34.097	27.069	4.454
2	7.692			10	22.538		
2	6.663			10	28.437		

Table II. Summary of three point flexure test results for heat treated samples.

Fiber Weight (%)	Heat Treatment (°C)	Max Strength (MPa)	Average Strength (MPa)	Standard Deviation (MPa)
2	200	5.943		
2	200	5.827		
2	200	5.927	5.438	0.523
2	200	4.965		
2	200	4.779		
2	200	5.186		

2	400	1.468		
2	400	1.582		
2	400	1.362	1.465	0.429
2	400	0.943		
2	400	2.214		
2	400	1.220		
2	600	0.966		
2	600	1.067		
2	600	0.730	0.888	0.122
2	600	0.864		
2	600	0.916		
2	600	0.786		
5	600	4.373		
5	600	4.003		
5	600	3.982	4.290	0.497
5	600	5.251		
5	600	3.950		
5	600	4.181		

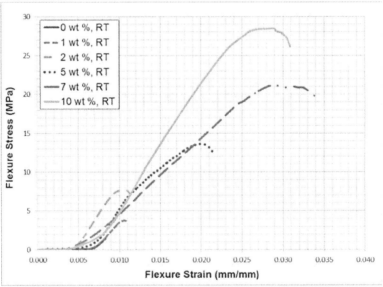

Figure 1. Representative stress-strain curves for room temperature samples.

Effect of Fiber Weight Percent

Figure 2 shows the average strength as a function of the amount of fiber reinforcement for all room temperature samples. Error ranges are based on one standard deviation, which was only calculated for the data in this study. The 1/4 inch fiber data are taken from Rill et al. for comparison purposes.[4] Table III summarizes the chart and includes calculations for the ratio, in percent, of the reinforced geopolymer strength to that of the un-reinforced geopolymer measured in each study. Strength increases at a slightly faster rate with increasing fiber weight fraction for the 1/2 inch long fibers compared to the 1/4 inch long fiber reinforcements. The addition of only one weight percent of fiber reinforcement resulted in more than twice the average flexure strength for both fiber lengths. At higher fiber concentrations, the most gains in strength are seen from the longer fibers. All batches showed some variability in maximum flexure stress values, which can be attributed to the hand-poured processing technique for the sample bars resulting in varying amounts of void content in the bars.

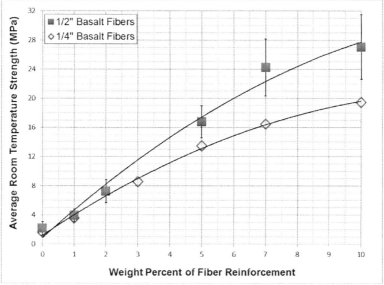

Figure 2. Average room temperature strength versus weight percent for 1/4 and 1/2 inch basalt fiber reinforcement.

Comparing the data for 1/4 inch long fiber reinforced geopolymer to that of the 1/2 inch fiber reinforced geopolymer for five weight percent or greater; 20-50% strength increases were seen when using the longer fibers. At fiber weight percent of five and above, hand-mixing was used instead of machine-mixing to avoid breaking up the fibers. A comparison between machine mixed and hand mixed flexure strength at five weight percent yielded a 20% increase in strength when the fibers are hand-mixed in. The average flexure strength for machine-mixed samples was 13.96 MPa compared to the 16.76 MPa shown above. In general higher weight percent of fiber reinforcement resulted in higher stress and strain at failure, which equates to higher strain energy and better fracture toughness.

Table III. Summary of average room temperature strength as a function of fiber weight percent for 1/4 and 1/2 inch long basalt fiber reinforcement.

Fiber Weight (%)	Fiber Length (in)	Average Strength (MPa)	Percent Baseline Strength (%)	Strength Increase (¼ → ½ inch) (%)
0	1/4	1.66	100.00	
0	1/2	2.16	100.00	
1	1/4	3.60	217.00	9.03
1	1/2	3.93	236.59	
2	1/2	7.26	437.37	
3	1/4	8.60	518.38	
5	1/4	13.50	813.74	24.13
5	1/2	16.76	777.63	
7	1/4	16.50	994.58	46.93
7	1/2	24.24	1124.97	
10	1/4	19.50	1175.41	38.82
10	1/2	27.07	1256.10	

Fracture surfaces showed large areas of planar fracture with both fiber fracture and fiber pullout visible without magnification. Fiber pullout is most prevalent in the room temperature samples. The fiber sizing may have evaporated at the elevated temperatures resulting in a strong fiber/matrix interphase allowing crack propagation from the matrix directly through the fibers. Fiber pullout can be clearly seen in Figure 3, which is an edge view of the fracture surface of a two weight percent, 200°C treated sample.

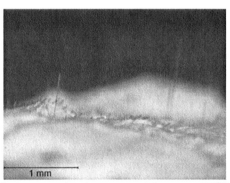

Figure 3. Edge-view of the fracture surface from a two weight percent, 200°C treated sample showing fiber pullout.

Residual Strength after Heat Treatment

Two and five fiber weight percent samples were heat treated prior to mechanical testing as discussed earlier. For the two fiber weight percent samples, heat treatments were performed at 200°C, 400°C and 600°C; whereas the five fiber weight percent samples were only heat treated at 600°C. Heat treatment was purposely kept below the leucite crystallization temperature of 1000°C, seen in potassium geopolymer by Bell et al.[6]. After being allowed to cool back to room temperature, the samples were tested in the same manner as the room temperature samples to determine the residual strength of the composite after heat treatment.

Figure 4 below shows the average residual strength values as a function of heat treatment temperature. The data are summarized in Table IV, which also shows the percent of baseline (not heat treated) strength that was retained after the respective heat treatment. On average, nearly 75% of the baseline strength is retained in the two fiber weight percent samples after heat treating to 200°C, but strength drops off significantly to 12% when heat treated to 600°C. The five fiber weight percent samples maintained 31% of room temperature strength after a heat treatment to 600°C.

Pure geopolymer, when heated, will shed water up to about 400°C. The water shed results in dehydration cracking of the material which increases in quantity and severity as heat treatment temperature increases. The addition of fiber reinforcement aided in reducing the severity of the dehydration cracking by bridging the cracks as they develop. High fiber concentrations also provide pathways for water to diffuse out of the samples when heated due to differences in thermal expansion coefficients or open space as a result of polymer sizing evaporation. For the two fiber weight percent samples in this study, at 200°C there is no evidence of dehydration cracking, but at the higher temperatures there are significant amounts of dehydration microcracking seen on the surfaces of the samples. The five fiber weight percent samples heated to 600°C showed dehydration cracking, but much less severe than that of the two fiber weight percent samples heated to the sample temperature. The dehydration cracking definitely contributed to the reduced strength of the composite bars. Figure 5 through Figure 8 below show the effect of dehydration on the heat treated samples as seen on a Leica Microsystems GmbH, model MZ6, Wetzlar, Germany, visible light microscope with a Spot Insight QE digital camera.

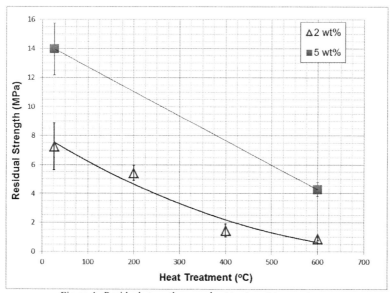

Figure 4. Residual strength versus heat treatment temperature.

Table IV. Summary of residual strength after heat treatment data.

Fiber Weight (%)	Heat Treatment (°C)	Average Strength (MPa)	Strength Retention (%)
2	26	7.26	100.00
2	200	5.44	74.94
2	400	1.47	20.19
2	600	0.89	12.24
5	26	13.96	100.00
5	600	4.29	30.73

Figure 5. Sample surface after heat treatment to 200°C, 2 wt % fiber reinforcement.

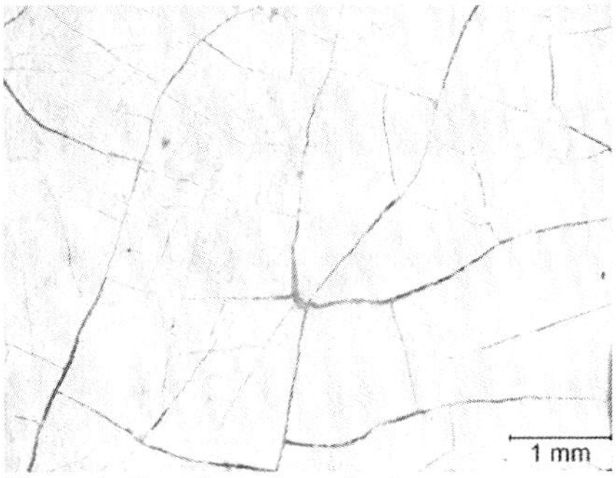

Figure 6. Sample surface after heat treatment to 400°C, 2 wt % fiber reinforcement.

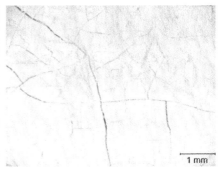

Figure 7. Sample surface after heat treatment to 600°C, 2 wt % fiber reinforcement.

Figure 8. Sample surface after heat treatment to 600°C, 5 wt % fiber reinforcement.

Average mass of the composite samples also decreased with increasing heat treatment temperature as summarized in Table V below and depicted along with the data for pure geopolymer from Rahier et al.[8] in Figure 9. Since specimen volume did not change noticeably as a result of heat treatment, the reduced mass also equated to lower sample density. As compared to the data for pure geopolymer[8], the samples with fiber reinforcement slowed the initial rate of water loss during heat treatment. However, by 600°C the free water was nevertheless completely removed.

Table V. Summary of average residual mass after heat treatment data.

Fiber Weight (%)	Heat Treatment (°C)	Average Mass (g)	Average Weight (%)
2	26	163.61	100.00
2	200	145.62	89.00
2	400	125.39	76.64
2	600	120.59	73.70
5	26	166.67	100.00
5	600	124.77	74.86

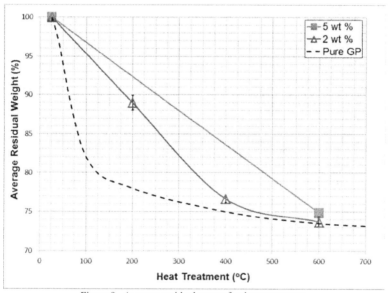

Figure 9. Average residual mass after heat treatment.

Weibull Statistical Analysis

Weibull statistics were also calculated using the maximum flexure stress data. A distribution function, F, was estimated based on the maximum failure stresses using the median rank method shown in equation (2). Where the maximum stress for each sample is given a rank, i, from 1 to n, which is the number of samples in the batch; with a rank of 1 given to the lowest stress value.[7] Data for pure geopolymer samples were provided by Rill et al.[4]

$$F \cong \frac{i-0.3}{n+0.4} \tag{2}$$

The Weibull distribution function is also given by equation (3) below, where v is the non-dimensional volume, σ is the stress, σ_o is the scale parameter and m is the shape parameter or Weibull modulus.

$$F = 1 - e^{\left[-v\left(\frac{\sigma}{\sigma_o}\right)^m\right]} \tag{3}$$

To determine the Weibull parameters, it was assumed that gage length is unchanged in the specimens, so $v = 1$, and equation (2) can be rearranged to get equation (3) below.

$$\ln\left[\ln\left(\frac{1}{1-F}\right)\right] = m \ln(\sigma) - m \ln(\sigma_o) \tag{4}$$

This fits the common slope-intercept form of a line, so $\ln\left[\ln\left(\frac{1}{1-F}\right)\right]$ was then plotted versus $\ln(\sigma)$, using the approximation of F given in equation (2), and a linear fit was applied for each data series. The slope of the fit line then corresponds to the Weibull modulus, m, and the y-intercept corresponds to $-m \ln(\sigma_o)$ which can be solved for σ_o. Table VI summarizes the resulting Weibull parameters for each sample batch. Figure 10 is the final Weibull function plot as a function of $\ln(\sigma)$.

The results of the room temperature Weibull statistical analysis show an increase in reliability over pure geopolymer due to the addition of fiber content. Heat treated samples showed no correlation between heat treatment temperature and reliability.

Table VI. Weibull parameters.

Fiber Weight (%)	Heat Treatment (°C)	m	σ_o (MPa)
0	RT	2.089	2.515
1	RT	4.793	4.281
2	RT	4.083	8.001
2	200	10.315	5.686
2	400	3.616	1.626
2	600	7.590	0.941
5	RT	7.844	14.774
5	600	7.834	4.550
7	RT	6.464	25.908
10	RT	6.136	29.027

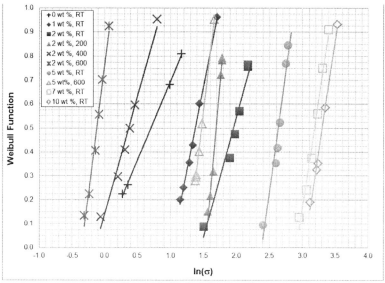

Figure 10. Weibull function plot.

CONCLUSIONS

Results of this study have shown that chopped basalt fiber reinforcement, when added to potassium geopolymer; significantly improve the room temperature, three point flexure strength. Strength improved from 2.16±0.94 MPa for pure geopolymer to 27.07±4.45 MPa with only ten weight percent fiber reinforcement. Basalt fiber reinforcement (two weight percent) also proved to preserve strength better after heat treatment, showing 75% strength retention after being heated to 200°C. Flexure strength continued to reduce as heat treatment temperature increased, but the composite still maintained 12% strength with two fiber weight percent and 31% strength with five fiber weight percent, when heated to 600°C. Potassium geopolymer reinforced with 1/2 inch long basalt fibers also showed 20-50% improved strength, at a weight percent greater than five, over those reinforced with 1/4 inch long basalt fibers for the same fiber weight percent.

Based on the void content of cured samples and variability in failure stresses for a given sample batch, further refinement of the processing technique would result in improved strength and less variance. At higher fiber concentrations (≥ 5), low shear hand mixing is a more effective method than machine mixing to avoid breaking up the longer fibers. Future work will include various and more advanced processing techniques, as well as other types and architectures of fiber reinforcement.

ACKNOWLEDGEMENTS

The authors thank the U.S. Air Force Office of Scientific Research (AFOSR), Tyndall Air Force Base, Florida for funding this study under grant AFOSR FA8650-11-1-5900 and the E.T. Horn Company. La Mirada, California and manufacturer Kamenny Vek Ltd., Moscow Region, Russia for supplying the samples of chopped basalt fibers.

REFERENCES

[1]"Understanding the Relationship Between Geopolymer Composition, Microstructure and Mechanical Properties" P. Duxson, J. Provis, G. Lukey, S. Mallicoat, W.M. Kriven, J.S. Deventer. Colloids and Surfaces A – Physicochemical and Engineering Aspects, **269** [1-3] 47-58 (2005).

[2]"Use of Geopolymeric Cements as a Refractory Adhesive for Metal and Ceramic Joins, J. L. Bell, M. Gordon and W. M. Kriven, Ceramic Engineering and Science Proceedings. Edited by D.-M. Zhu, K. Plucknett and W. M. Kriven, **26** [3] 407-413 (2005).

[3]"Geopolymers as Adhesives for Polymeric Foam," W.M. Kriven, B. Glad. Presented at PacRim9, July 2011.

[4]"Properties of Basalt Fiber Reinforced Geopolymer Composites," E. Rill, D.R. Lowry, W.M. Kriven. Ceramic Engineering and Science Proceedings, **31** (10) , pp. 57-67 (2010).

[5]"Fabrication of Structural Leucite Glass-Ceramics from Potassium-Based Geopolymer Precursors" Ming Xie, Lonathan L. Bell, and Waltraud M. Kriven. J. Amer. Ceram. Soc., **93** [9] 2644-2649 (2010).

[6]"Formation of Ceramics from Metakaolin-based Geopolymers: Part II. K-based Geopolymer," J. L. Bell, P. E. Driemeyer and W. M. Kriven, J. Amer. Ceram. Soc., **92** [3] 607-615 (2009).

[7]"Statistical Analysis of Fracture Strength of Composite Materials Using Weibull Distribution," M. Hüsnü Dirikolu, Alaattin Aktas and Burak Birgören, Turkish J. Eng. Env. Sci., **26,** 45-48 (2002).

[8]"Low-temperature synthesized aluminosilicate glasses: Part II: Rheological transformations during low-temperature cure and high-temperature properties of a model compound ," Rahier, H., B. VanMele, and J. Wastiels. Journal of Materials Science, **31** 80-85, (1996)

[9]"Effect of High Tensile Strength Polypropylene Chopped Fiber Reinforcements on the Mechanical Properties of Sodium Based Geopolymer Systems," D. R. Lowry and W. M. Kriven. Ceramic Engineering and Science Proceedings, **31** (10), (2010).

[10]"Chemical Composition and Mechanical Properties of Basalt and Glass Fibers: A Comparison," T. Deak and T. Czigany. Textile Research Journal, **79** (7) (2009).

[11]"Synthesis and Thermal Behaviour of Potassium Sialate Geopolymers," Valeria F.F. Barbosa and Kenneth J.D. MacKenzie. Materials Letters, **57**, 1477– 1482 (2003).

CERAMICASH: A NEW ULTRA LOW COST CHEMICALLY BONDED CERAMIC
MATERIAL

Henry A. Colorado[1, 2*], Jenn-Ming Yang[1]

[1]Materials Science and Engineering, University of California, Los Angeles.
[2]Universidad de Antioquia, Mechanical Engineering Department. Medellin, Colombia.

ABSTRACT

CeramicAsh is a new concept of green materials fabricated mainly from fly ash, reacting with an acidic liquid such as phosphoric acid formulations. These materials, depending on the ratio of the constituents, can vary from a glazy dense to a micro or nano porous ultra-low density structure. Microstructure was characterized using a scanning electron microscope and X-ray diffraction. Density measurement and compression tests were conducted for samples with different compositions. Results show *CeramicAsh* is an ultra-light ceramic with high compressive strength and resistant to high temperatures. They are promising materials for thermal, structural and hazardous waste stabilization applications.

INTRODUCTION

High temperature manufacturing processes contribute to global warming and this contribution is especially significant in the processing of cementitious and ceramic materials. Each year, coal-burning power plants, steel factories and similar facilities in the United States produce more than 125 million tons of waste, much of it fly and bottom ash left from combustion. This quantity depends on the fuel type, raw ingredients used and the energy efficiency of the cement plant [1].

Conventional ceramics usually require high temperature during a part of the manufacturing process, which is undesirable because it increases cost and has a negative environmental impact. Sintered ceramics have been used for thousands of years by humans and even today are the subject of a very intense research field mainly at high temperatures. However, sintering requires a lot of energy and the process is expensive at large manufacturing scales. The solution is chemical bonding, like in Portland cement (although making this cement requires a lot of energy during the complete process), which allows this product to be inexpensive in high volume production. Ceramics are expensive compared to cement but they have in general higher mechanical strength, corrosion resistance and temperature stability. There are several emerging materials which have properties in between: Chemically Bonded Ceramics (CBCs), which have been developed intensively during the last decade to fill the gap between cements and ceramics.

Chemically bonded ceramics (CBCs) have been extensively used for multiple applications because they combine thermomechanical properties of sintered ceramics with the easy fabrication of conventional cements [2-4]. Their applications include: radioactive and hazardous waste treatment (nuclear waste solidification and encapsulation) [5-6], shielding different types of radiation [3], dental cements [7], patch repair material [8], bone tissue engineering [9], high temperature [10], composites with fillers and reinforcements [3, 7, 11-16], electronic materials [17], tooling for advanced composites [18], and coatings on nanotubes and nanowires applications [19]. The Chemically Bonded Phosphate Ceramics (CBPCs) can reach a compressive strength of over 100 MPa in minutes after fabrication, whereas Portland cement based concrete reaches a compressive strength about 20MPa after 28 days. CBPC's density is usually below 2.2 g/cm^3. This is particularly interesting when compared with

43

Portland cement, which is generally around 3.15 g/cm^3. This opens up many applications related to infrastructure repair such as roads, bridges and pipes, as well as other structural applications such as ceramic composite firewalls, towers, turbines and aerospace applications.

CBCs are a class of materials that is classified as Acid-Base cements, which involves acid-base and hydration reactions. A complete overview of the field of Acid Base-Cements can be reviewed in [20]. These materials bridge the gap between the attributes of sintered ceramics and traditional hydraulic cements. Often there is a need for materials with properties in between these two and CBCs fill this need [3]. CBCs have mechanical properties approaching to sintered ceramics [2] and high stability in acidic and high temperature environments; its processing is inexpensive, castable, and environmentally friendly [4]. The bonding in such CBPCs is a mixture of ionic, covalent, and van der Waals bonding, with the ionic and covalent dominating. In traditional cement hydration products, van der Waals and hydrogen bonding dominate [2].

In this paper, fly ash is used as the source of the ceramic composite itself, which represent an enormous decrease in costs and weight not only with respect to conventional ceramics and cement, but also with respect to previous formulations of CBPCs [3] and other geopolymers. This new material uses more waste ashes than normal CBPCs, which themselves used to be a big source of contamination, thus reducing their impact in the environment. On the other hand, CeramicAsh has shown good compression strength, which enables this new material as a substitute for cementituos materials. Additionally, porosity can be controlled by different methods including mixing time, particle size and liquid to powder ratio. Because of the origin and nature of the fly ashes, this new composite material opens up many applications including fire resistant walls and coatings, structural composites and electronics materials when the corresponding additives are added.

EXPERIMENTAL

The manufacturing of CBPC samples was conducted by mixing an aqueous phosphoric acid formulation, fly ash and in some cases natural wollastonite as reinforcement. The pH of the CBPC after curing was near neutral in all cases. The composition of fly ash (from Diversified Minerals Inc) and wollastonite (from Minera Nyco) are presented in Tables 1 and 2 respectively.

Table 1. Chemical composition range of fly ash class F

Composition	CaO	SiO_2	Fe_2O_3	Al_2O_3
Weight %	5-22	59-63	2-5	11-15

Table 2. Chemical composition of wollastonite powder.

Composition	CaO	SiO_2	Fe_2O_3	Al_2O_3	MnO	MgO	TiO_2	K_2O
Weight %	46.25	52.00	0.25	0.40	0.025	0.50	0.025	0.15

For compression samples, 2.0, 1.2, 1.0 and 0.5 phosphoric acid formulation to fly fsh were fabricated. In addition, samples with wollastonite ($CaSiO_3$) contents were fabricated in order to see its effect of the mechanical properties.

A Teflon® fluoropolymer mold with mold release (Synlube 1000 silicone-based release agent applied before the mixture discharge) was used to minimize the adhesion of the CBPC to the mold. Next, the mold with the CBPC was covered with plastic foil to prevent exposure to humidity and decrease shrinkage effects. Samples were released after 48 hours and then dried at room temperature in open air for at least 3 days. Samples were then mechanically polished with parallel and smooth faces (top and bottom) for the compression test. Since the CBPC has both unbonded and bonded water, samples were dried slowly in the furnace in order to prevent residual stresses first at 50˚C for 1 day, followed by 105˚C for an additional day.

Compression tests were conducted in an Instron® machine 3382, over cylindrical CBPC samples (9 mm in diameter by 20 mm in length) for M200, M200 with 1, 10, 20 and 50% of fly ash. In addition, the effect of the mixing time on the compression strength was studied for the CBPC with 1% of fly ash. A set of 20 samples were tested for each powder. The crosshead speed was 1mm/min.

To see the microstructure, sample sections were ground using silicon carbide papers of 500, 1000, 2400 and 4000 grit progressively. After polishing, samples were dried in a furnace at 70˚C for 4 hours Next, samples were mounted on an aluminum stub and sputtered in a Hummer 6.2 system (15mA AC for 30 sec) creating a 1nm thick film of Au. The SEM used was a JEOL JSM 6700R in the high vacuum mode. Elemental distribution x-ray maps were collected on the SEM equipped with an energy-dispersive analyzer (SEM-EDS). The images were collected on the polished and gold-coated samples, with a counting time of 51.2 ms/pixel. X-Ray Diffraction (XRD) experiments were conducted usin X'Pert PRO equipment (Cu Kα radiation, λ=1.5406 Å), at 45Kv and scanning between 10° and 70°. M200 and M200 with 1, 10, 20 and 50% of fly ash samples were ground in an alumina mortar and XRD tests were conducted at room temperature.

Finally, density tests were conducted over CBPCs with fly ash as filler. All samples were tested after a drying process (50˚C for 1 day, followed by 100˚C for 1 day) in a Metter Toledo[TM] balance, by means of the buoyancy method. Six samples for each composition were tested. The dry weight (Wd), submerged weight (Ws), and saturated weight (Wss) were measured. The following parameters were calculated:

Bulk volume: $Vb = Wss - Ws$; Apparent volume: $Vapp = Wd - Ws$; Open-pore volume: $Vop = Wss - Wd$; % porosity = $(Vop/Vb) \times 100$ %; Bulk Density: $Db = Wd/(Wss - Ws)$; and Apparent Density: $Da = Wd/(Wd - Ws)$. In these calculations the density of water was taken to be 1.0 g/cm^3.

RESULTS

Figure 1 (a, b and c) show SEM images of fly ash class C, fly ash class F and CaSiO$_3$, which correspond to the solid raw materials (as received) used in this research. Figure 1 (d) shows a cross section view image of a fabricated CBPC with fly ash.

The pH results for 2.0, 1.5, 1.0 and 0.5 phosphoric acid formulation to fly ash ratio (wt%/wt%) were respectively 7, 6, 5 and 7. In all cases after the liquid mixture completed to set, the pH was about 7.

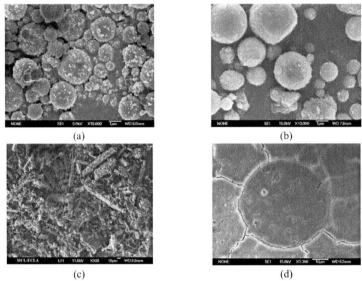

Figure 1. Fly ash class C (a), fly ash class F (b), CaSiO₃ (c), and CBPC with fly ash (d).

XRD data presented in Figure 2 shows new crystalline phases grown as a result of the acid base reaction, which has been identified as calcium phosphates. These phases are calcium, iron and aluminum phosphates and combinations of them. Amorphous phases of those are also present. The role of all these phases on the composite need to be further investigated.

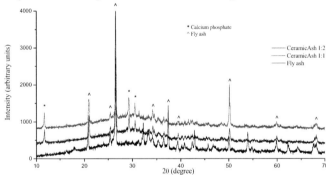

Figure 2. XRD spectra for different materials fabricated based only in fly ash.

Figure 3 shows different SEM images for the new material based on fly ash. Images were obtained from fractured exposed surfaces or from as-grown surfaces.

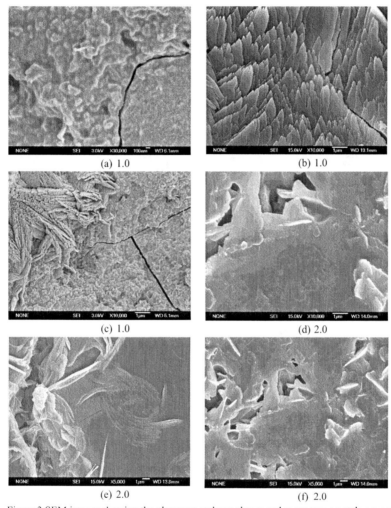

(a) 1.0

(b) 1.0

(c) 1.0

(d) 2.0

(e) 2.0

(f) 2.0

Figure 3 SEM images showing the glassy amorphous phases and some new crystals grown during the chemical reaction at low temperature for CeramicAsh. The acidic formulation to fly ash (wt%/wt%) is indicated.

Figures 3 (a, b and c) show images for samples with 1.0 phosphoric acid formulation to fly ash ratio (wt%/wt%). Figure 3 (a) shows the matrix, mainly containing amorphous phases. Different new grown phosphate phases appear like in Figures 3 (b and c). A more glassy material was obtained for 2.0 phosphoric acid formulation to fly ash ratio, as shown in Figures 3 (d, e and f).

Figures 4a and b show images for samples with 1.5 phosphoric acid formulation to fly ash ratio. Figure 4 (a) shows a fly ash particle completely bonded to the ceramic matrix as the result of the reaction with the phosphoric acid formulation. Figure 4 (b) shows phosphates growing in this glazy material. Figures (c and d) show images for samples with 0.5 acidic formulations to fly ash ratios (wt%/wt%). This material has more pores and the matrix is not glazy which looks like a traditional cement matrix.

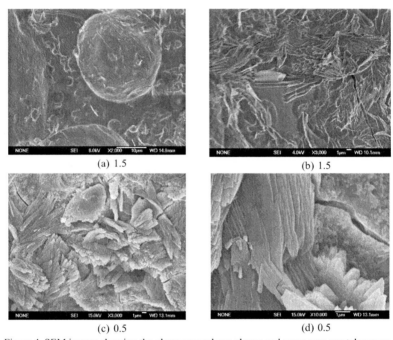

(a) 1.5

(b) 1.5

(c) 0.5

(d) 0.5

Figure 4 SEM images showing the glassy amorphous phases and some new crystals grown during the chemical reaction at low temperature for CeramicAsh. The acidic formulation to fly ash is indicated.

Some properties of the material presented above are summarized on Table 3. As the acidic formulation to fly ash ratio increases, current values for compressive strength and density were found to increase. The compressive strength is under improvement and some manufacturing procedures followed in the recently developed wollastonite-based Chemically Bonded Phosphate ceramics [21-23] will be applied.

Table 3 Properties for CeramicAsh without reinforcement

Acidic formulation to fly ash ratio	Compressive strength [MPa]	Density [g/cm³]
2.0	12.6	2.2
1.5	7.0	1.9
1.0	12.0	1.9
0.5	3.0	1.4

Table 4 and their corresponding EDS images, Figure 5, shows the element contents for a sample with 1.0 phosphoric acid formulation to fly ash ratio.

Table 4 EDS results for images shown on Figure 5.

Element wt%	B	O	Na	Mg	Al	Si	P	Tl	Ca	V	Eu	Gd	Fe	Yb
Figure 5a	0.00	34.91	3.45	1.70	5.09	0.32	34.14	5.52	5.84	0.26	2.09	1.69	2.35	2.62
Figure 5b	27.87	18.25	1.42	0.87	2.96	0.87	25.15	4.60	6.86	0.59	1.18	2.61	1.59	5.17
Figure 5c	0.00	2.84	0.17	0.19	2.29	0.74	35.00	1.62	21.63	1.83	6.90	8.88	7.64	10.27

The amount of phases generated and their interrelation in final properties of the composite material needs to be further investigated and it shows the potential the material has for different applications that requires specific chemistry, pH and different properties.

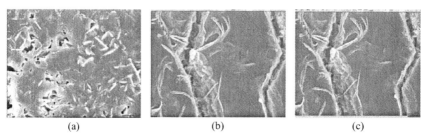
(a) (b) (c)
Figure 5 SEM images showing the spots were the EDS was conducted for samples with 1.0 phosphoric acid formulation to fly ash ratio.

On the other hand, when wollastonite (CaSiO₃) is added to CeramicAsh as reinforcement, there is an increase in compressive strength. It is associated with the role that calcium has in producing new strong amorphous binding phases. Figure 6 shows the XRD for fly ash, the material fabricated with M200 wollastonite powder and CBPCs with fly ash. Figure 6a shows that in general fly ash is difficult to identify in the CBPC, with only slight differences in intensity, even with 50 wt% as filler. This can be explained from composition results presented in

Table 2, in which the first four constituents of wollastonite (CaO, SiO$_2$, Fe$_2$O$_3$, Al$_2$O$_3$), which make up 98.9% of the material, are also the major contents present in the fly ash. This means that almost the same products can be obtained when these powders react with the phosphoric acid. Also, it is observed that mixing time does not have significant effects on the composition of the composite.

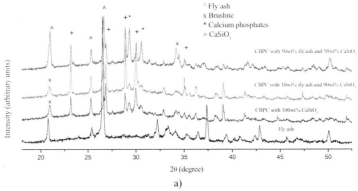

a)

Figure 6 XRD for fly ash, CBPC and CBPCs with fly ash as filler.

Figure 7 shows the Weibull distributions for the CeramicAsh with CaSiO$_3$ as reinforcement. The largest values were obtained for the CBPC with no filler.

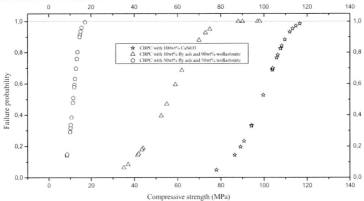

Figure 7 Weibull distributions for CBPCs fabricated with wollastonite powder M200 and with fly ash.

DISCUSSION

In this paper, fly ash is used as the source of the ceramic composite itself, which represents an enormous decrease in cost and weight with respect to both conventional ceramics and cement as well as previous formulations of CBPCs and other geopolymers. Because of the mechanical properties presented above, we are proposing that there should be further investigation into this material. The results can produce not only a good solution for replacing high performance cement concrete but also a high performance material to be used in firewalls and aerospace applications.

Potential applications

This material is fabricated at room temperature with very little processing in order to replace high performance concrete cement. Since we are using fly ash contaminants as the main component and very little amount of phosphoric acid in a very inexpensive acidic solution, we are not only proposing a competitive inexpensive material but also a good solution for the environment.
Some of the potential applications for the proposed material are:

i) Fire resistant and retardant coatings for wood and other materials. The thermal properties for CeramicAsh are similar to fly ash since this is the main component. Since fly ash is composed of micro spheres, thermal insulation is a very promising field for CeramicAsh. Since pH can be controlled to neutral, inexpensive E-glass fibers can be used, unlike traditional Portland cement where the pH is in excess of 11 and only acid resistant glass fibers can be used. Other fibers also can be easily incorporated to produce high performance beams as we did in pultrusion for Wollastonite based Chemically Bonded Phosphate Ceramics [22]. CeramicAsh is a very inexpensive ceramic when it is compared with traditional ceramics that involve high temperature processing, or even when it is compared with other emerging chemically bonded ceramics that involve more expensive metal oxides.

ii) Alternative materials as a replacement of gypsum in building applications, with highest fire resistant properties. In CBPCs, chemistry can be tailored easily to obtain materials from acidic to alkaline pH.

iii) Composite micro and nano capacitors: micro hollow spheres in a ceramic matrix composite have shown to have good dielectric properties.

iv) Storage of nuclear wastes (in compositions where ash compositions lead to immobilization of wastes and the consolidation of a solid material): since the ceramics are fabricated by chemistry; both the pH and final chemical composition can be easily tailored to produce specific solutions depending on the nuclear waste that need to be treated.

v) High temperature resistant paint. Since CBPCs start from a liquid solution that hardens with time [2], CeramicAsh can be applied as paint when the appropriate additives are incorporated in the acidic formulation and the suitable manufacturing process is applied to decrease the viscosity of the mixture. In addition, the setting time can be extended from minutes to hours.

vi) Semi-transparent porous glass applications (can be tailored as a glass porous material obtained at room temperature). Different colors are possible.

vii) Aerospace applications since this new material have excellent fire resistant properties and very low density as well.

ACKNOWLEDGEMENTS
The authors wish to thank to Colciencias from Colombia for the grant Laspau-Fulbright to Henry A. Colorado.

REFERENCES
[1] Environmental Protection Agency AP 42 - Compilation of Air Pollutant Emission Factors, Volume I *Stationary Point and Area Sources*, Arunington, DC, 2005.

[2] Della M. Roy. New Strong Cement Materials: Chemically Bonded Ceramics, Science, Vol. **235**: 651-58 (1987).

[3] A. S. Wagh. Chemical bonded phosphate ceramics, *Elsevier,* Argonne National Laboratory, USA, 283, 2004.

[4] S. Y. Jeong and A. S. Wagh. Chemical bonding phosphate ceramics: cementing the gap between ceramics, cements, and polymers, Argonne National Laboratory report, 2002.

[5] S. Chattopadhyay. Evaluation of chemically bonded phosphate ceramics for mercury stabilization of a mixed synthetic waste. National Risk Management Research Lab. Cincinnati, Ohio, 2003.

[6] D. Singh, S. Y. Jeong, K. Dwyer and T. Abesadze. Ceramicrete: a novel ceramic packaging system for spent-fuel transport and storage. Argonne National Laboratory.Proceedings of Waste Management 2K Conference, Tucson, AZ, 2000.

[7] L. C. Chow and E. D. Eanes. Octacalcium phosphate. Monographs in oral science, vol 18. Karger, Switzerland, 2001.

[8] Fei Qiao, C.K. Chau, Zongjin Li. Property evaluation of magnesium phosphate cement mortar as patch repair material. Construction and Building Materials, **24** 695–700, (2010).

[9] S. Barinov and V. Komlev. Calcium Phosphate based bioceramics for bone tissue engineering. Trans Tech Publications Ltd, Switzerland, 2008.

[10] M. A. Gulgun, B. R. Johnson and W. M. Kriven. Chemically bonded ceramics as an alternative to high temperature composite processing. Mat. Res. Soc. Symp. Proc. Vol. 346. Materials Research Society, (1994).

[11] T. L. Laufenberg, M. Aro, A. Wagh, J. E. Winandy, P. Donahue, S. Weitner and J. Aue. Phosphate-bonded ceramic-wood composites. Ninth International Conference on Inorganic bonded composite materials, 2004.

[12] J. Lin, S. Zhang, T. Chen, C. Liu, S. Lin and X. Tian. Calcium phosphate cement reinforced by polypeptide copolymers. J Biomed Mater Res Part B, 76B 432–439, (2006).

[13] K. Park and T.Vasilos. Characteristics of carbon fibre-reinforced calcium phosphate composites fabricated by hot pressing. Journal of Materials Science Letters **16** 985–987, (1997).

[14] J. Pera and J. Ambroise. Fiber-reinforced magnesia-phosphate cement composites for rapid repair. Cement and Concrete Composites **20** 31-39, (1998).

[15] L. A., Dos Santos, L. C. Oliveira, E. C. Silva, R. G. Carrodeguas, A. Ortega and A. C. Fonseca. Fiber reinforced calcium phosphate cement. 24 **3** 212-216, Blackwell Science, Inc. International Soc. Artificial Organs, (2000).

[16] D. Hong-Lian, W. Xin-Yu, H. Jian, Y. Yu-hua and L. Shi-Pu. Effect of carbon fiber on calcium phosphate bone cement. Trans. Nonferrous Met. Soc. China. Vol **14** N.4, 769-774, (2004).

[17] J. F. Young and S. Dimitry. Electrical properties of chemical bonded ceramic insulators. J. Am. Ceram. Soc., **73**: 9 2775-78, (1990).

[18] L. Miller and S. Wise. Chemical bonded ceramic tooling for advanced composites. Materials and Manufacturing Processes, Volume 5, Issue 2, 229-252, (1990).

[19] A. Gomathi, S. R. C. Vivekchand, A. Govindaraj and C. N. Rao. Chemically bonded ceramic oxide coatings on carbon nanotubes and inorganic nnowires. Adv. Mater., 17, 2757-2761, (2005).

[20] Wilson, A. D. and Nicholson, J. W. Acid based cements: their biomedical and industrial applications. Cambridge, England, Cambridge University Press, 1993.

[21] Henry A. Colorado, Clem, Hiel, Jenn-M. Yang, H. Thomas Hahn. Wollastonite-based chemically bonded phosphate ceramic composites. Metal, ceramic and polymeric composites for various uses. ISBN 979-953-307-135-9, (2011).

[22] H. A. Colorado, H. T. Hahn and C. Hiel. Pultrusion of glass and carbon fibers reinforced Chemically Bonded Phosphate Ceramics. Journal of Composites Materials. Vol 45 no. 23 2391-2399, (2011).

[23] H. A. Colorado, C. Hiel and H. T. Hahn. Chemically Bonded Phosphate Ceramics composites reinforced with graphite nanoplatelets. Composites Part A. Composites: Part A 42 376–384, (2011).

CHEMICALLY BONDED PHOSPHATE CERAMICS FOR STABILIZATION OF HIGH-SODIUM CONTAINING WASTE STREAMS

H. A. Colorado, Roopa Ganga, Dileep Singh
Nuclear Engineering Division, Argonne National Laboratory. Argonne, IL 60439

ABSTRACT
Chemically bonded phosphate ceramics (CBPC) to stabilize high sodium containing waste streams were fabricated and tested in this work. CBPCs based waste forms were prepared by mixing the sodium-based simulant, with magnesium oxide (MgO) and Class C fly ash as filler. A mechanical mixer was used for mixing all the components. Different compositions and samples sizes were fabricated. Microstructure was identified with X-ray diffraction and electron microscopy. Compressive strength was also evaluated for the compositions in which a solid material was obtained. A set of 33 different chemical compositions have been fabricated in this project and only few of them did not set into a solid material. A maximum concentration of 60% of $-NaH_2PO_4$ was attained in the set product.

INTRODUCTION

This paper presents Chemically Bonded Phosphate Ceramics (CBPC) as a solution to stabilize high sodium containing wastes. The main contribution in this research is the solidification of these solutions that come from the nuclear waste processing step with very high concentrations of sodium phosphates.

Typically, sodium wastes are the product of nuclear site operations such as decontamination activities, some of which use dilute sodium hydroxide to wash surfaces and solubilize residues. As a result, significant sodium-based salts are present in the liquid wastes. The high sodium content in these solutions makes them unsuitable for direct calcination as   the sodium nitrate melts at low temperatures and does not produce a granular, free-flowing calcination product.

Since these high content sodium salts wastes are very sensitive to high temperatures, an ideal solution is to stabilize them at room temperature in solid materials like ceramics. It is well known that ceramics usually require high temperature processing like sintering. These processes are not desirable because they involve a lot of energy consumption and also are not cost effective for large-scale manufacturing. Their use of coal as a fuel also leads to an increase in emitted gases and particles causing these processes to have one of the worst environmental impacts in the manufacturing industry. It is well established that about 1000 kg of CO_2 is emitted for every 1000 kg of Portland cement produced in the U.S. The quantity depends on the fuel type, raw ingredients used and the energy efficiency of the cement plant [1].

Fortunately, there is an emerging technology to produce ceramics fabricated by chemical reactions, Chemically Bonded Phosphate Ceramics (CBPCs), which are environmentally benign and fabricated at room temperature. Like in Portland cement (although making Portland cement requires a lot of thermal energy when the complete process is considered) the bonding is obtained by chemistry, which allows this product to be inexpensive in high volume production [2]. Chemically bonded ceramics [3] refers to the bonding that takes place in a chemical reaction at low temperatures, as opposed to thermal diffusion or sintering at temperatures (even higher than 1000°C) in traditional ceramics and cements. In traditional cement hydration products, van der Waals and hydrogen bonding dominate [4]. The bonding in such CBCs is a mixture of ionic,

55

covalent, and van der Waals bonding, with the ionic and covalent dominating. In addition, CBCs possess excellent fire resistance and thermal insulation; manufacturing is easy, inexpensive and environmentally benign, and can be cast in any shape.

The CBPCs form by acid-base reactions between an acid phosphate (such as that of potassium, ammonium or aluminum) and a metal oxide (such as that of magnesium, calcium, or zinc). Comprehensive and complete reviews are available about the different CBPCs developed ([1],[3-8].

There are diverse applications for CBPCs such as dental material [9] and for bone tissue engineering [10], radiation shielding [11], nuclear waste solidification and encapsulation [12], electronics [13], advanced composites [14], coatings on nanotubes and nanowires applications [15] and as an alternative to high temperature composites processing [16]. In addition, CBPC with fillers and reinforcements like CBPC-wood composites [17], CBPC-glass fibers reinforcements [3] and CBPC-fly ash [1] have also been produced.

The CBPCs presented in this research are the result of mixing high-sodium containing waste streams (fabricated with different concentrations of anhydrous monobasic sodium phosphate (NaH_2PO_4), water, fly ash and magnesium oxide (MgO)).

The waste steam formulation was made of different concentrations of NaH_2PO_4 dissolved in water. Temperature, mixing times and other parameters were extensively explored in order to have high loadings of NaH_2PO_4 in the final ceramic. These liquid formulations utilized were selected to simulate high sodium content wastes typically generated at nuclear power plants. A typical waste formulation preparation consisted of dissolving 400 g of anhydrous monobasic sodium phosphate in 600 mL of water. The mixture was stirred and subsequently, MgO and fly ash components were added to make the final slurry which was poured in molds for setting.

A maximum concentration of 60% of NaH_2PO_4 was consolidated into a solid ceramic material, which is very high loading for this type of materials. These high loadings of a sodium-based liquid solution into a ceramic material fabricated at room temperature qualify this solution as the best in our understanding. Further, the procedure presented is economical and environmentally friendly. Composition analysis of the samples show that the samples are composites of residual phases from the materials used as raw materials, and also from new crystalline and non-crystalline phases formed during the reaction that acts as the binding phase. The new phases were magnesium and magnesium-sodium phosphates, such as $MgHPO_4.6H_2O$ and $MgNaPO_4.7H_2O$. The amorphous magnesium and sodium phases appear mixed in sort of a continuous amorphous phases matrix.

EXPERIMENTS

The liquid formulations utilized were selected to simulate high sodium content wastes typically generated at nuclear power plants. Anhydrous monobasic sodium phosphate (NaH_2PO_4) was used as a source for sodium. Sodium phosphate monobasic dehydrate ($NaH_2PO_4 \cdot 2H_2O$) was used in some cases for comparison with NaH_2PO_4 in the waste simulant fabrication. A typical waste formulation preparation consisted of dissolving 400 g of anhydrous monobasic sodium phosphate in 600 mL of water. The mixture was stirred for 15 minutes with a magnetic stirrer until all particles were completely dissolved. For the higher loadings of sodium, the solution was heated up to 60°C while being continuously stirred with a magnetic stirrer in order to completely dissolve the monobasic sodium phosphate. Temperature of the solution was then gradually reduced while continuous stirring to avoid precipitation. Subsequently, MgO and fly ash components were added to make the final slurry which was poured in molds for setting. Two cylindrical sample sizes: 20 mm diameter x 30 mm length and 50 mm diameter x 100 mm length were fabricated using plastic molds.

After setting for at least 2 weeks, the samples were retrieved from the molds for characterizations. A diamond saw was used to cut cylinders with parallel and smooth flat surfaces. The samples were then dried for one day in the fume hood at room temperature. Compression tests were conducted in an Instron machine 3382. A set of 3 samples were tested for each composition at a crosshead speed of 1mm/min.

Compression tests were conducted for waste streams with 13.9, 19.5, 21.9, 29.3, 30, 35, 35.6, 50 and 55 wt% of NaH_2PO_4. Different mixing times, chemical compositions and variations of the manufacturing procedures were tested in order to find the best conditions in which the components set into a solid ceramic material. Raw materials, depending on the size and composition, were mixed manually or mechanically.

X-Ray Diffraction (XRD) experiments were conducted using an X'Pert PRO (Cu Kα radiation, λ=1.5406 Å), at 45KV and scanning between 10° and 80°. Density was measured by weighing the samples just before the tests and calculating the volume from measurements taken on each sample.

For microstructural evaluation, sample sections were ground using silicon carbide papers grit ANSI 240, 400 and 1200 incrementally, and then they were polished with alumina powders of 1μm grain size. After polishing, samples were dried in open air for 3 days. For SEM examination, samples were mounted on an aluminum stub and sputtered in a Hummer 6.2 system (15mA AC for 30 sec), creating approximately a 1 nm thick film of Au. Scanning electron microscope (SEM) used was a JEOL JSM 6700R in high vacuum mode. Elemental distribution x-ray maps were collected on the SEM with an energy-dispersive X-ray spectroscopy analyzer (SEM-EDS), with a counting time of 51.2 ms/pixel. In some samples, fracture surfaces were observed, for which the gold coating was also included.

ANALYSIS AND RESULTS
Figure 1 shows SEM images of MgO, NaH_2PO_4 and Fly Ash Class C, which correspond to the solid raw materials as received used in this research.

a) MgO   b) NaH_2PO_4

c) Fly Ash type C

Figure 1. Raw materials used in the fabrication of the high-sodium containing waste streams.

Typical images for mechanical and hand-made samples are presented in Figure 2. Temperature of the solution was always measured during the mixing process, using a thermocouple placed 1cm below the liquid surface, in order to keep the slurry temperature < 60°C. For some samples, dry ice was used below the container to cool the slurry. Thus, by controlling the slurry temperature, setting time is increased and the heat released during the exothermic reaction (that takes place during the mixing) is reduced. These combined effects produce an increase in the mixing time, which is ideal not only for the manufacturing process but also for the control of the chemical reaction of the products and phases generated in the set product. The ideal mixing time was established to be about 20 min. This gives enough time to properly mix the components and to have them react and consolidate into a stable solid material.

Figure 2. a) mechanical mixing (for S9 sample), b) Typical sample (S9) after 1 week of manufacturing, c) typical hand mixing (for S17 sample), and d) sample (S17) after 1 week of manufacturing.

Figure 3 shows images corresponding to the samples fabricated classified by the NaH_2PO_4 contents. NS indicates that the samples did not set up properly in a hard solid material. Table 1 shows details of the main components used for each composition fabricated in weight percentage.

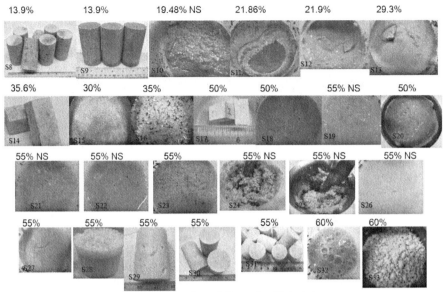

Figure 3. Overview of all samples fabricated.

Table 1. Compositional make-up of the various samples fabricated (in wt.%).

Sample	NaH₂PO₄	MgO	water	Fly Ash	Sample	NaH₂PO₄	MgO	water	Fly Ash
S1	9.7	66.0	23.7	0.0	S18	50.0	3.4	35.7	10.9
S2	6.8	66.0	25.2	0.0	S19	55.0	5.7	39.3	0.0
S3	10.0	33.0	24.0	33.0	S20	50.0	10.9	35.7	3.4
S4	10.0	33.0	24.0	33.0	S21	55.0	5.0	39.3	0.7
S5	10.0	33.0	24.0	33.0	S22	55.0	5.0	39.3	0.7
S6	10.0	32.7	23.8	32.7	S23	55.0	9.0	34.0	2.0
S7	10.0	32.7	23.8	29.4	S24	55.0	11.0	34.0	0.0
S8	13.9	33.0	19.1	33.0	S25	55.0	11.0	34.0	0.0
S9	13.9	33.0	19.1	33.0	S26	55.0	8.5	36.5	0.0
S10	19.5	8.7	46.3	25.6	S27	55.0	11.0	34.0	0.0
S11	21.9	9.7	39.7	28.7	S28	55.0	10.0	35.0	0.0
S12	21.9	9.7	39.7	28.7	S29	55.0	11.0	34.0	0.0
S13	29.3	8.8	35.9	26.0	S30	55.0	11.0	34.0	0.0
S14	35.6	5.6	42.5	16.4	S31	55.0	11.0	34.0	0.0
S15	30.0	8.4	36.7	24.9	S32	60.0	10.0	30.0	0.0
S16	35.0	5.7	42.7	16.7	S33	60.0	10.0	30.0	0.0
S17	50.0	10.9	35.7	3.4					

Figure 4 shows NaH₂PO₄wt% for all samples fabricated in this project.

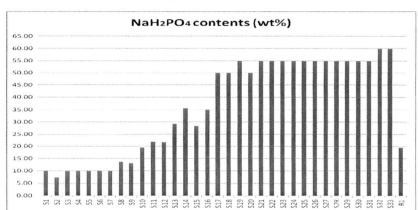

Figure 4. NaH$_2$PO$_4$ for different samples fabricated.

Figure 5 shows MgO/NaH₂PO₄ mol ratio for all samples fabricated in this project.

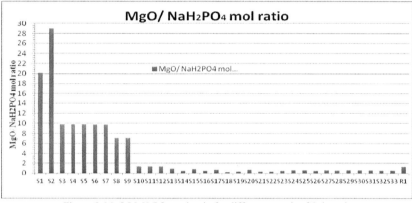

Figure 5. MgO/NaH$_2$PO$_4$ mol ratio for different samples fabricated.

Figure 6 shows the compressive strength for some of the representative samples fabricated. As shown, almost all samples fulfill the requirement for materials used in nuclear waste stabilization of 3.45MPa (500psi). However, it was found that some of them set very slowly and that the compressive strength increased over time in general for all samples.

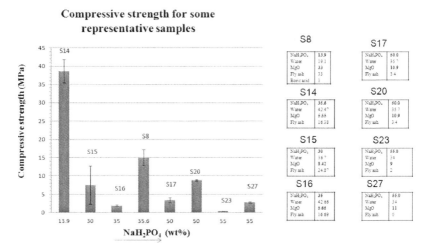

Figure 6 Compressive strength for some representative samples. Their respective raw materials formulation is shown.

Figure 7 shows XRD patterns for some of the representative samples fabricated. MgO, Fly Ash, NaH₂PO₄ and the new phosphates phases are shown with different symbols. These results show that some most of the samples fabricated have remaining phases from the raw materials used during the synthesis. In addition, new crystalline phases are shown, mostly associated to two groups: magnesium and magnesium-sodium phosphates.

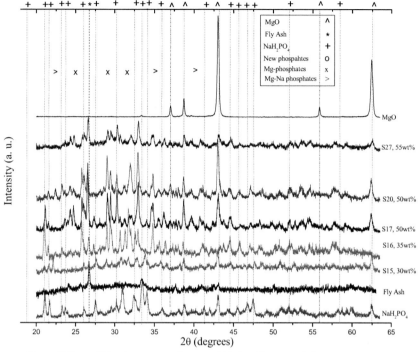

Figure 7 X ray diffraction patterns for several high-sodium containing waste streams.

Figure 8 shows SEM images for fractured samples. Figure 8a shows a sample with 14wt % NaH_2PO_4, some Fly Ash particles included. Figure 8b shows a sample with 35wt % NaH_2PO_4, with some new grown phosphate phases are shown, Figure 8c shows a sample with 55wt % NaH_2PO_4 with remaining NaH_2PO_4 grains shown, and finally Figure 8d shows a sample with 60wt % NaH_2PO_4.

Figure 9 shows X ray maps for the sample with 14wt % NaH_2PO_4 (S8). Qualitative maps are shown for P, Na, Mg, Al, Si and Ca. All of them were taken over the area of Figure 9a.

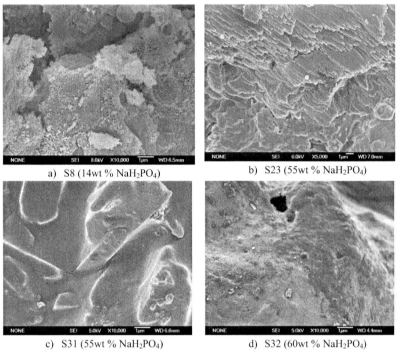

a) S8 (14wt % NaH_2PO_4)

b) S23 (55wt % NaH_2PO_4)

c) S31 (55wt % NaH_2PO_4)

d) S32 (60wt % NaH_2PO_4)

Figure 8 SEM images for some of the samples fabricated.

EDAX experiments were conducted over the same area and the corresponding spectrum is shown in Figure 9i.

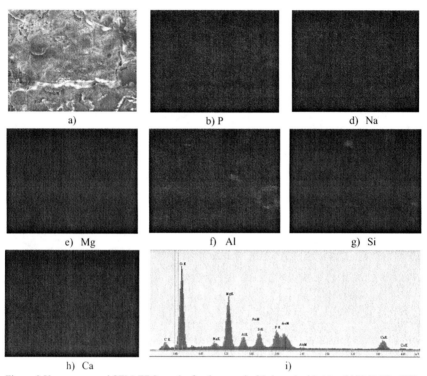

a)   b) P   d) Na

e) Mg   f) Al   g) Si

h) Ca   i)

Figure 9 X ray maps and SEM-EDS results for the sample fabricated with 14wt % NaH$_2$PO$_4$ (S8).

The quantification of the EDS is shown in Table 2.

Table 2 Quantitative SEM-EDS for sample the fabricated with 14wt % NaH$_2$PO$_4$ (S8).

Element	C	O	Ja	Mg	Al	Si	P	Ca	Au
Wt%	8.93	33.68	1.35	13.23	3.34	5.12	7.45	9.28	17.61
At%	17.22	48.74	1.36	12.60	2.87	4.22	5.57	5.36	2.07

Figure 10 shows a ternary diagram of the samples manufactured. Samples that did not set are also shown.

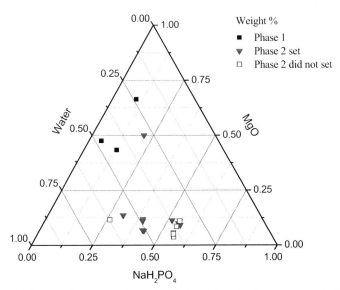

Figure 10. Ternary diagram for all samples fabricated showing the components in Weight percentage.

DISCUSSION

In this research, a CBPC was fabricated with NaH_2PO_4, MgO, water and (in some cases) with Fly Ash class C. Results show that the samples are composites of residual phases from the materials used as raw materials, and also from new crystalline and non-crystalline phases formed during the reaction that acts as the binding phase. The new phases were shown in the XRD and they were associated to two groups: magnesium and magnesium-sodium phosphates, such as $MgHPO_4.6H_2O$ and $MgNaPO_4.7H_2O$. A representation of the phases in the samples fabricated containing all the presented raw materials is shown in Figure 11. The amorphous magnesium and sodium phases appear mixed in sort of a continuous matrix. From the XRD patterns, as the NaH_2PO_4 concentration increased, the relative amount of amorphous phases increased and the crystalline phases decreased.

From all shown phases, the amorphous material seem to be the most significant binding structure in the fabricated ceramic. Also, it was observed that with time, samples show an aging effect. This was reflected on the increase in the compressive with time and not necessarily from the water loss (less than 5%).

As a future work, a study of the aging effect on the microstructure for longer times than 3 months will be conducted to establish the long-term behavior of the phosphate material and its suitability in disposal environments.

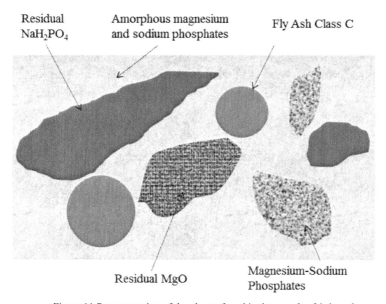

Figure 11 Representation of the phases found in the samples fabricated.

SUMMARY
Chemically bonded phosphate ceramics for stabilizations of high-sodium containing waste streams have been successfully fabricated in this research through combinations of anhydrous monobasic sodium phosphate (NaH_2PO_4), water , Fly Ash and magnesium phosphate (MgO). Up to a maximum concentration of 60 wt.% of NaH_2PO_4 in the fabricated ceramics. A set of 33 different chemical compositions have been fabricated in this project.

These high loadings of a sodium-based liquid solution in a ceramic material fabricated at room temperature qualify this CBPC solution as a potential solution for the treatment of sodium wastes. The procedure shown using hand and mechanical mixing are economical and environmentally friendly than other solutions that involve high temperature processing.

Also, it was shown that the sodium-based CBPC is a composite material itself, in which the diverse phases generated can be tailored for other applications. Further experiments will be conducted to see the role of the phases on the mechanical, thermal and chemical properties of these sodium based ceramic matrix composites.

ACKNOWLEDGMENTS
Authors want to thank the Japan Atomic Energy Agency (JAEA) for the support as part of a Work-for-Others project.

REFERENCES
[1]. Environmental Protection Agency AP 42 - Compilation of Air Pollutant Emission Factors, 2005 Volume I *Stationary Point and Area Sources*, Arunington, DC
[2]. S. Y. Jeong and A. S. Wagh. Chemical bonding phosphate ceramics: cementing the gap between ceramics, cements, and polymers. Argonne National Laboratory report, June 2002.
[3]. A. S. Wagh. Chemical bonded phosphate ceramics. 2004 *Elsevier* Argonne National Laboratory, USA. 283.
[4]. Della M. Roy. New Strong Cement Materials: Chemically Bonded Ceramics. February 1987 *Science*, Vol. **235** 651
[5]. Wilson, A. D. and Nicholson, J. W. Acid based cements: their biomedical and industrial applications. Cambridge, England, Cambridge University Press, 1993.
[6]. Arun S. Wagh and Seung Y. Jeong. Chemically bonded phosphate ceramics: I, a dissolution model of formation. J. Am. Ceram. Soc., 86 [11] 1838-44 (2003).
[7]. Arun S. Wagh and Seung Y. Jeong. Chemically bonded phosphate ceramics: II, warm-temperature process for alumina ceramics. J. Am. Ceram. Soc., 86 [11] 1845-49 (2003).
[8]. Arun S. Wagh and Seung Y. Jeong. Chemically bonded phosphate ceramics: III, reduction mechanism and its application to iron phosphate ceramics. J. Am. Soc., 86 [11] 1850-55 (2003).
[9]. L. C. Chow and E. D. Eanes. Octacalcium phosphate. Monographs in oral science, vol 18. Karger, Switzerland, 2001.
[10].   Sergey Barinov and Vladimir Komlev. Calcium Phosphate based bioceramics for bone tissue engineering. Trans Tech Publications Ltd, Switzerland, 2008.
[11].   S. Chattopadhyay. Evaluation of chemically bonded phosphate ceramics for mercury stabilization of a mixed synthetic waste. National Risk Management Research Lab. Cincinnati, Ohio, 2003.
[12].   D. Singh, S. Y. Jeong, K. Dwyer and T. Abesadze. Ceramicrete: a novel ceramic packaging system for spent-fuel transport and storage. Argonne National Laboratory.Proceedings of Waste Management 2K Conference, Tucson, AZ, 2000.
[13].   J. F. Young and S. Dimitry. Electrical properties of chemical bonded ceramic insulators. J. Am. Ceram. Soc., 73 [9] 2775-78 (1990).
[14].   L. Miller and S. Wise. Chemical bonded ceramic tooling for advanced composites. Materials and Manufacturing Processes, Volume 5, Issue 2 1990, 229-252.
[15].   A. Gomathi, S. R. C. Vivekchand, A. Govindaraj and C. N. Rao. Chemically bonded ceramic oxide coatings on carbon nanotubes and inorganic nnowires. Adv. Mater. 2005, 17, 2757-2761.
[16].   M. A. Gulgun, B. R. Johnson and W. M. Kriven. Mat. Res. Soc. Symp. Proc. Vol. 346. Materials Research Society, 1994.
[17].   T. L. Laufenberg, M. Aro, A. Wagh, J. E. Winandy, P. Donahue, S. Weitner and J. Aue. Phosphate-bonded ceramic-wood composites. Ninth International Conference on Inorganic bonded composite materials. October 2004.

Computational Design,
Modeling, and Simulation

NUMERICAL SIMULATION OF THE TEMPERATURE AND STRESS FIELD EVOLUTION APPLIED TO SPARK PLASMA SINTERING

J.B. Allen*, C. Walters[+], C.R. Welch*, and J.F. Peters*
*U.S. Army Engineer Research and Development Center, Vicksburg, MS 39180-6199
[+]Department of Materials, Imperial College London, London SW7 2AZ U.K.

ABSTRACT
Spark plasma sintering is a high-amperage, low-voltage powder consolidation technique that employs pulsed direct current and uniaxial pressure. Over the past several years, it has been successfully used to produce a variety of different materials including metals, composites, and ceramics. This work presents a transient finite element model of aluminum oxide sintering that incorporates a coupled electrical, thermal, mechanical analysis that closely resembles the procedures used in physical experiments. Within this context, this paper outlines the governing equations that pertain to a balanced energy equation and includes the effects of thermal and electrical contact forces, radiation, and Joule heating. These equations are coupled with the relevant equations pertaining to mechanical displacements and prescribe the necessary initial and boundary conditions for a complete solution. As part of this transient analysis, the implementation of a Proportional Integral Derivative controller, which affords the use of a predetermined heating rate conditioned upon a variable voltage, is also presented. Finally, implications relating to the temperature and stress fields are discussed, and possible avenues for improvement are suggested.

INTRODUCTION

Spark plasma sintering (SPS) is a high-amperage, low-voltage powder consolidation technique that employs pulsed direct current (DC) and uniaxial pressure.[1-4] Over the past several years, SPS has been successfully used to produce a variety of different materials including metals,[5] composites,[6] and ceramics.[7,8] In contrast to the more traditional (pressure assisted) methods such as hot pressing (HP) or hot isostatic pressing (HIP), a significant benefit of using SPS includes the reduction of processing times and sintering temperatures. This helps to minimize grain growth and thereby result in mechanical,[9] physical,[10] and optical[11] improvements in the material.

Over the past decade, SPS research has increased significantly, as evidenced by the increased number of papers on the subject.[3] While a significant proportion of this research activity has been strictly experimental, in recent years the number of numerical simulation studies has increased. Insights gained from these numerical simulations have helped to overcome some of the inherent challenges associated with performing experimental observations such as the direct determination of the temperature and stress field evolution within the sample. Further, these simulations have in general led to a better understanding of the cause and effect relationships between the evolving processing parameters (i.e., stress and temperature fields) and the evolution of the material microstructure.

Several recent works have used the finite element method to investigate the SPS current and temperature distributions within the sample.[12-19] While each of these studies has played a significant role in enhancing the overall understanding of the SPS process, there is still considerable room for improvement. Relatively few of these studies, for example, take into account the presence of contact resistances,[13-16] and all but three[17-19] omit the contribution from thermal and applied stresses. This latter omission is particularly relevant since several experimental studies have demonstrated that the inclusion of processing stresses is of vital importance to densification mechanisms.[3, 20]

This work presents a transient finite element model that incorporates a coupled electrical, thermal, mechanical analysis that closely resembles the actual SPS experimental conditions. Within this context, the paper outlines the governing equations that pertain to a balanced energy equation and includes the effects of contact forces, radiation and Joule heating. This information is coupled with the relevant equations pertaining to mechanical displacements, and the necessary initial and boundary conditions for a complete solution are prescribed. As part of this transient analysis, the implementation of a Proportional Integral Derivative (PID) controller is presented, which, similar to actual experimental conditions, affords the use of a predetermined heating rate conditioned upon a variable voltage. Attention is focused on two candidate materials, fully dense alumina (Al_2O_3), and fully dense silicon carbide (SiC), and experimental validation on each of these is provided. Finally, implications relating to the temperature and stress fields are discussed, and possible avenues for improvement are suggested.

GEOMETRY

Figure 1 shows a schematic of the SPS apparatus (SPS SYNTEX 2050) used in this work. Processing is conducted within a vacuum chamber and consists of the sample, a graphite die, two punches, six graphite spacers, and two steel electrodes, each located at the extreme top and bottom of the setup. In addition to providing a potential difference for electrical current flow, the electrodes are also subject to a compressive mechanical load. As shown, the sample (diameter 19 mm, height 3 mm) is surrounded radially by a graphite die (thickness 12.75 mm, height 38 mm)   and is compressed longitudinally by two opposing punches (height 25 mm). Other relevant dimensions are shown in Figure 1.   Points A and B, corresponding to the upper sample surface, are shown for reference purposes.

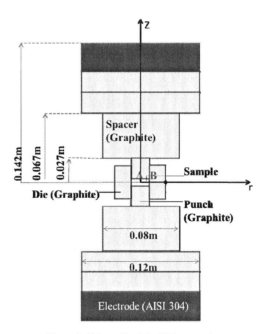

Figure 1. Schematic of the SPS apparatus

GOVERNING EQUATIONS

The set of equations governing the three-way, coupled interaction between the electrical, thermal, and mechanical system are shown in Equations 1-10. The four dependent variables (degrees of freedom)—temperature, the radial and axial components of displacement, and the electric potential—are solved through an energy balance approach acting over the entire apparatus. As indicated in Equation 1, the relative heat gain by conduction is governed by Fourier's law, while the remaining heat transfer mechanisms, q_{conv}, q_j, q_{rad}, and q_{cont} corresponding to convection, Joule heating, radiation, and contact resistance, respectively, are assigned to source terms. In addition to the contribution from uniaxial loading, the mechanical stress, shown in Equations 8-10, is also coupled to the energy equation, through thermal expansion. Additionally, since the majority of the material properties are thermally dependent (see Table I), this requires an iterative approach to convergence and an update of these material properties over each successive time-step.

Performing an energy balance on the SPS apparatus produces the following equation:

$$\rho C_p \frac{\partial T}{\partial t} = \nabla \cdot (k \nabla T) + q_j + q_{rad} + q_{conv} + q_{cont} \tag{1}$$

where

$\rho =$ sample density
$C_p =$ sample specific heat at constant pressure
$T =$ temperature
$k =$ thermal conductivity

and q_j, q_{rad}, q_{conv}, and q_{cont} are source terms corresponding to Joule heating, radiative heat transfer, convective heat transfer, and contributions from thermal and electrical contact resistances, respectively.

Since the experiments take place within a vacuum, the heat loss due to convection was neglected ($q_{conv} = 0$). Surface to ambient radiation occurs among the exposed surfaces of the die, spacers, and punches, and is given by:

$$q_{rad} = \sigma_s A_s \varepsilon (T_e^4 - T_a^4) \tag{2}$$

where

$\sigma_s =$ Stefan-Boltzmann constant ($\sigma_s = 5.67 \cdot 10^{-8}$ W/m^{2}K^4)
$A_s =$ area of the exposed surface
$\varepsilon =$ emissivity ($\varepsilon_{graphite} = 0.8$ as given by Zavaliangos et al.[12] and Vanmeensel et al.[13])
$T_e, T_a =$ emitting and absorbing surface temperatures, respectively

The term q_j represents the rate of Joule heating per unit volume and is given by

$$q_j = \mathbf{J} \cdot \mathbf{E} \tag{3}$$

where $\mathbf{J}$ is the current density and $\mathbf{E}$ is the electric field. These in turn are computed via solutions of Equations (4) and (5), where σ is the electrical conductivity and φ is the electric potential. As shown, the electrostatic condition is imposed to reduce modeling complexity, and is in general a valid assumption since the electric potential reaches a state of equilibrium far in advance of the temperature.

$$\nabla \cdot \boldsymbol{J} = 0 \tag{4}$$

$$\nabla \cdot \boldsymbol{J} = \nabla \cdot (\sigma \boldsymbol{E}) = \nabla \cdot (-\sigma \nabla \varphi) = 0 \tag{5}$$

It has been demonstrated that thermal and electrical resistances induced at both the horizontal and vertical contacts between the various components of the tool assembly and sample can have a significant influence on the evolution of the temperature and electrical potential fields.[12,13] While several different approaches exist to account for these resistances, this work treats them in accordance with Vanmeensel et al.[13]

$$q_{cont[th]} = h_g \Delta T \tag{6}$$

$$q_{cont[el]} = \sigma_g (\Delta V)^2 \tag{7}$$

where h_g and σ_g are the thermal and electrical contact conductance coefficients, respectively. The values for the horizontal ($h_{g|H}$ and $\sigma_{g|H}$) and vertical ($h_{g|V}$ and $\sigma_{g|V}$) coefficients used in this study were taken from "best-fit" approximations made by other authors[12,13], resulting in: $h_{g|H} = 15 \text{ E3 W/m}^2 \text{ K}$, $\sigma_{g|H} = 5 \text{ E7 } \Omega^{-1}\text{m}^{-2}$, $h_{g|V} = h_{g|H}/6$, and $\sigma_{g|H} = \sigma_{g|H}/6$.

Subsequent to the solution to the temperature field, the mechanical displacements are computed from the equations of linear elasticity, under the simplifying assumption that the sample yield stress is not exceeded. These are shown in Equations (8) through (10):

$$\nabla \cdot \boldsymbol{\sigma} + \boldsymbol{F} = 0 \tag{8}$$

$$\varepsilon = \frac{1}{2\mu}[\sigma - \nu(tr\sigma)I] + \alpha_T \nabla T I \tag{9}$$

$$\varepsilon = 0.5[\nabla \boldsymbol{u} + (\nabla \boldsymbol{u})^T] \tag{10}$$

where

- $\boldsymbol{\sigma} =$ the stress tensor
- $\boldsymbol{F} =$ body force per unit volume
- $\alpha_T =$ thermal coefficient of expansion
- $\varepsilon =$ strain tensor
- $\mu =$ Lame coefficient [$\mu = \mu(E, \nu)$] where E is Young's modulus
- $\nu =$ Poisson's ratio
- $\boldsymbol{u} =$ displacement vector

MESH AND SOLUTION PROCEDURE

The geometric symmetry of the SPS apparatus allows for a two-dimensional (2D), axisymmetric modeling approach in cylindrical coordinates (r, z). As such, the linearized partial differential equations were simplified to reflect this change of coordinates. The 2D domain was meshed via Delaunay triangulation and was composed of 12,428 three-node triangular elements with an average element quality of 0.95. For each node, four degrees of freedom were assigned: displacement in the r- and z-directions, temperature, and electric potential. The total number of degrees of freedom was 101,747. The numerical integration over each triangular element was approximated by the sum over all Gauss-point contributions and solved for each degree of freedom with an absolute error tolerance of 0.001 using the multiphysics simulation software COMSOL 4.1. Time-stepping was performed using a second-order backward differencing scheme.

MATERIAL PROPERTIES

The temperature-dependent material properties used in this study for the steel electrodes (AISI 304), the graphite tooling assembly and the Al_2O_3 sample are shown in Table I. The temperature dependent values shown correspond to temperatures between approximately 273 K and 2500 K. Apart from the temperature, density dependence is another important consideration for SPS analysis. Since the density of the graphite over this temperature range varies by less than 5 percent,[21] it was assumed constant and modeled as an isotropic material. In contrast, the Al_2O_3 is known to have a highly variable density that is dependent on the sintering stage.[20] This density dependence corresponds to the various diffusion-related processes that contribute to such densification mechanisms as grain growth, pore shrinkage, and pore annihilation. While various mesoscale modeling efforts, including Tikare et al.,[22] have made considerable progress in predicting this density evolution, still, the vast majority of numerical models are highly dependent on physical experimentation for their initial and boundary conditions. These limitations are further compounded at larger scales, where more of the apparatus is modeled. For these reasons, this work, like most previous studies,[12-19] assumes a sample of constant density, indicative of the final stages of sintering.

Table I. Material properties

	Graphite[1]	AISI 304[2]	Al_2O_3[2]
Density ρ (kg m^{-3})	1900	7900	3970
Specific Heat C_p ($J \cdot kg^{-1} \cdot K^{-1}$)	34.27 + 2.72T - 9.6 10^{-4}T^2	446.5 + 0.162T	-126.5317 + 8.1918T - 6.1058·10^{-3}T^2
Thermal Conductivity κ ($W \cdot m^{-1} \cdot K^{-1}$)	82.85 - 0.06T + 2.58·10^{-5}T^2	9.988 + 0.01746T	76.4488 - 0.18978T + 1.9596·10^{-4}T^2
Electric Resistivity ρe (Ω·m)	2.14·10^{-5} - 1.34 10^{-8}T + 4.42 10^{-12}T^2	10^{-8}[50.1685 + 0.0838T - 1.7453·10^{-5}T^2]	10^{-8}
Young's Modulus E (Pa)	103·10^9	200·10^9	215·10^9
Poisson's Ratio ν	0.32	0.33	0.32
Thermal Expansion Coefficient α (K^{-1})	8.0·10^{-6}	18·10^{-6}	8·10^{-6}

BOUNDARY CONDITIONS

As stated previously, all heat loss by convection was neglected. At the lower electrode boundary the following conditions were imposed: (i) a constant temperature of 300 K (due to refrigerant cooling); (ii) a ground state electric potential (V = 0); and (iii) a zero mechanical displacement condition. A surface to ambient radiation boundary was assigned to each of the graphite sides in accordance with Equation 2. The top electrode was assigned: (i) a constant temperature condition of 300 K; (ii) a constant applied load of 11.5 KN; and (iii) a time-varying voltage that functioned in accordance with a PID controller.

PID CONTROLLER

Thermally controlled SPS experiments follow a preset heating schedule that is input by the operator and is regulated by a time-changing current or voltage profile. This allows for a high degree of temperature control and thus better overall control over the evolving sample microstructure. Ideally, a numerical model should follow a similar control process. In fact, however, the majority of the numerical work done previously has been conducted under the assumption of a constant applied current.[11, 17, 19, 21] In the present work a PID controller was utilized for this purpose.

In general, a PID controller approximates an error correction between a measured process variable and a desired setpoint by the calculation and subsequent output of a corrective action, which can adjust the process in an efficient manner and maintain minimal error.[20] The PID controller thus works in an iterative, closed-loop configuration. In this work the applied voltage $V(t)$ is adjusted in accordance with a prescribed heating rate. A tracking error $e(t)$, representing the difference between the desired heating rate and the actual heating rate, is computed for a specified control point (see Point 1 of Figure 1). This error signal $e(t)$ is then sent to the PID controller, which computes the new voltage via the derivative and the integral of this error signal according to the following:

$$V(t) = K_p \cdot e(t) + K_i \cdot \int_0^t e(t)d\tau + K_D \frac{de(t)}{dt} \qquad (11)$$

where K_p, K_I, and K_D represent the proportional, integral, and derivative gains, respectively. The constant values for these gains were chosen for an optimal control response, and were assigned values in accordance with the Zeigler-Nichols method.[23]

RESULTS

The experiments were performed by controlling the temperature according to a predetermined time-temperature profile. As indicated in Figure 2, the time evolution of temperature involves a constant heating rate of 200 K/min for the first 300 sec, followed by a 300-sec dwell period holding at a temperature of 1300 K. Thereafter, at 600 sec, cooling is initiated as the applied loading and electrical power are terminated. The total duration of the simulated experiments is 900 sec (15 min). Figure 2 shows the target temperature profile (dashed line) and the actual simulated temperature (solid line) at point B. As indicated, the effectiveness of the PID controller allows for virtually an error-free temperature difference between the simulated and experimental results.

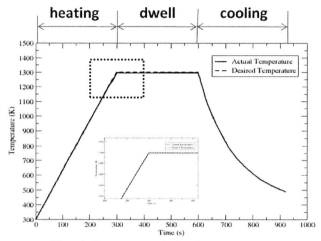

Figure 2. Temperature evolution via PID controller.

The corresponding evolution of the applied voltage potential is shown in Figure 3. As shown, after a nonlinear increase in the voltage during the heating ramp, a drop is observed at time $t = 300$ sec in order to stabilize the controlled temperature at 1300 K. From 300 to 600 sec, a slight decrease of the voltage potential is observed in order to compensate for the gradual heating of the large ram cylinders. Finally, at a time of 600 sec, the voltage is shut off to allow for the cool-down period.

Figure 4 shows the time evolution of the first invariant of the stress tensor (I_1) at point B. This tensor is computed from the principal stresses ($I_1 = \sigma_1 + \sigma_2 + \sigma_3$). As shown, after a slight change from tensile to compressive stress is encountered during the heating phase, a compressive stress of 30 MPa is maintained for the duration of the dwell period. Afterward, at the initiation of cooling (t=600 sec.), a sharp stress gradient occurs, reaching magnitudes approaching 100 MPa. This sharp stress gradient which is manifest for certain ceramic materials at the onset of the cooling phase could potentially lead to material defects and may warrant a controlled, more gradual cooling approach.

Figure 5 shows the temperature contours corresponding to the apparatus at $t = 300$ sec, corresponding to the end of the ramped heating stage. A maximum temperature of 1345 K was obtained within the central locations of the punches, while the sample temperature averaged approximately 1295 K, or approximately 55 percent of its melting temperature.

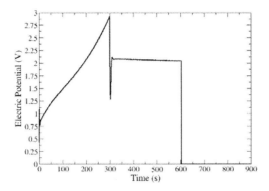

Figure 3. Time evolution of the electric potential at Point B

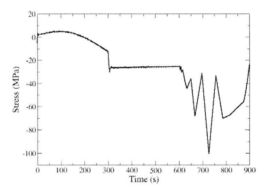

Figure 4. Time evolution of the first stress invariant at Point B.

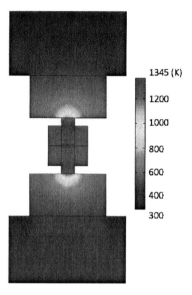

Figure 5. Temperature contours at time $t = 300$ sec for a heating rate of 200K/min.

The temperature contours in the sample/punch/die regions for the alumina sample at $t = 300$ sec and $t = 600$ sec (initiation of the cooling stage) for a heating rate of 200 K/min are shown in Figure 6(a) and Figure 6(b), respectively. As indicated, the temperature difference in the entire configuration was reduced because of the thermal equilibrium achieved during the dwell period. In agreement with Figure 5, the temperatures at the central cores of the punch exhibit the highest values, especially at the level close to the top free surface of die wall.

Within the alumina sample (Figure 7), a temperature gradient is observed in the radial direction. As indicated, the temperature difference between the sample core and the sample-die interface is approximately 17 K. As stated by Munoz and Anselmi-Tamburini,[20] this significant level of thermal nonhomogeneity may be attributed to the relatively low thermal conductivity of aluminum oxide, compared to other, more conductive sample specimens, which tend to reveal a significantly more uniform temperature distribution for equivalent operating conditions. Like the non-uniform stress distributions, and depending on their magnitude, these nonhomogeneous temperatures within the sample could potentially affect the inner microstructure of the material (i.e., nonuniform grain sizes), and result in material defects.

The stress distribution (I_1), plotted at time t=350 s, and as a function of radial distance within the sample, is shown in Figure 8. Like the temperature, the compressive stress is also non-homogeneous, exhibiting a stress variation of approximately 15 MPa between points A and B.

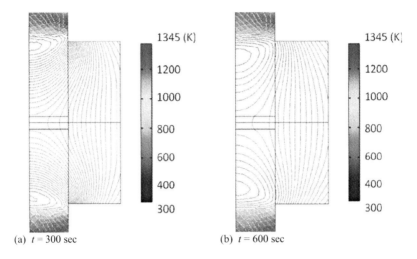

(a) $t = 300$ sec   (b) $t = 600$ sec

Figure 6. Temperature surface contour plot for a heating rate of 200 K/min.

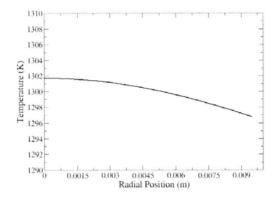

Figure 7. Temperature distribution inside the alumina sample at
$t = 350$ sec. and at line AB

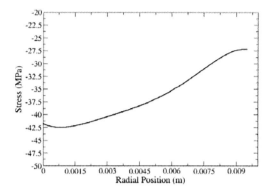

Figure 8. Stress distribution (first invariant) inside the alumina sample at
$t = 350$ sec. and at line AB

CONCLUSIONS
    In this work we presented a transient, finite element heat transfer model of aluminum oxide sintering appropriate to the SPS process. The model utilized a coupled approach employing an electrical, thermal, and mechanical analysis which included the use of a PID controller. This controller, which adjusted the applied voltage to agree with a preset heating schedule, proved particularly valuable for this type of thermally controlled SPS processes. Comparisons of the model with physical experiments revealed excellent agreement in terms of quantifying the temperature evolution at a fixed point outside of the sample. Inside the sample, the numerical results showed significant, nonhomogeneous temperatures and stresses occurring between the sample core and the sample die interface. These temperatures and stresses ranged from 17 K and 15 MPa, respectively. It was noted, that these nonhomogeneous temperature and stress fields could adversely affect the quality of the interior microstructure and produce undesirable results.
    Because of the limiting assumption relating to a constant density sample, the results presented here may be fully applicable only at the final stages of sintering, wherein the sample microstructure is fully dense and considered isotropic. In order to expand the utility of the model, future implementations will necessarily include the effects of sample densification.

ACKNOWLEDGMENTS
This study was funded through the U.S. Army Engineer Research and Development Center Directed Research Program Advanced Material Initiative.

REFERENCES

[1]J. R. Groza, Field Assisted Sintering, *ASM Handbook*, **7**, 583-589 (1998).
[2]R. Orru, R. Licheri, A. Locci, A. Cincotti, and G. Cao, Consolidation/Synthesis of Materials by Electric Current Activated/Assisted Sintering, *Mat. Sci. Eng. R: Reports*, **63**, 127–287 (2009).

[3]Z. A. Munir, U. Anselmi-Tamburini, and M. Ohyanagi, The Effect of Electric Field and Pressure on the Synthesis and Consolidation of Materials: A Review of the Spark Plasma Sintering Method, *Journal of Materials Science*, **41,** 763-777 (2006).

[4]R.V. Lenel, Resistance Sintering under Pressure, *Transactions AIME*, **203,** 158-167 (1955).

[5]V. Y. Kodash, J. R. Groza, K. C. Cho, B. R. Klotz, and R. J. Dowding, Field-Assisted Sintering of Ni Nanopowders, *Mater. Sci. Eng. A*, **385,** 367-371 (2004).

[6]Z. Shen, M. Johnsson, Z. Zhao, and M. Nygren, Spark Plasma Sintering of Alumina, *J. Am. Ceram. Soc.*, **85,** 1921-1927 (2002).

[7]R. Kumar, K. H. Prakash, P. Cheang, and K. A. Khor, Microstructure and Mechanical Properties of Spark Plasma Sintered Zirconia-Hydroxyapatite Nano-Composite Powders, *Acta Materialia*, **53,** 2327-2335 (2005).

[8]K. Morita, B. N. Kim, K. Hiraga, and H. Yoshida, Fabrication of Transparent $MgAl_2O_4$ Spinel Polycrystal by Spark Plasma Sintering Process, *Scripta Materialia*, **58,** 1114-1117 (2008).

[9]Z. Shen, Z. Zhao, H. Peng, and M. Nygren, Formation of Tough Interlocking Microstructures in Silicon Nitride Ceramics by Dynamic Ripening, *ļature* , **417,** 266-269 (2002).

[10]K. A. Khor, K. H. Cheng, G. L. Yu, and F. Boey, Thermal Conductivity and Dielectric Constant of Spark Plasma Sintered Aluminum Nitride, *Mater. Sci. Eng. A*, **347,** 300-305 (2003).

[11]U. Anselmi-Tamburini, S. Gennari, J. E. Garay, and Z. A. Munir, Fundamental Investigations on the Spark Plasma Sintering/Synthesis Process II. Modeling of Current and Temperature Distributions, *Mater. Sci Eng A*, **394,** 139-148 (2005).

[12]A. Zavaliangos, J. Zhang, M. Krammer, and J. R. Groza, Temperature Evolution during Field Activated Sintering, *Mater Sci Eng A*, **379,** 218-228 (2004).

[13]K. Vanmeensel, A. Laptev, J. Hennicke, J. Vleugels, and O. Van der Biest, Modeling of the Temperature Distribution during Field Assisted Sintering, *Acta Materialia*, **53,** 4379-4388 (2005).

[14]B. McWilliams, A. Zavaliangos, K. C. Cho, and R. J. Dowding, The Modeling of Electric-Current-Assisted Sintering to Produce Bulk Nanocrystalline Tungsten, *JOM*, **58**(4), 67–71(2006).

[15]B. McWilliams, and A. Zavaliangos, Multi-Phenomena Simulation of Electric Field Assisted Sintering, *J. Mater Sci.*, **43**(14), 5031–5503 (2008).

[16]G. Antou, G. Mathieu, G. Trolliard, and A. Maitre, Spark Plasma Sintering of Zirconium Carbide and Oxycarbide, *J. Mater Res*, **24**(2), 404–412 (2009).

[17]X. Wang, S. R. Casolco, G. Xu, and J. E. Garay, Finite Element Modeling of Electric Current Activated Sintering: The Effect of Coupled Electrical Potential, Temperature and Stress, *Acta Materialia*, **55,** 3611–3622 (2007).

[18]U. Anselmi-Tamburini, J. E. Garay, and Z. A. Munir, Fast Low-Temperature Consolidation of Bulk Nanometric Ceramic Materials, *Scripta Materialia*, **54,** 823–828 (2006).

[19]G. Maizza, S. Grasso, and Y. Sakka, Moving Finite-Element Mesh Model for Aiding Spark Plasma Sintering in Current Control Mode of Pure Ultrafine WC Powder, *J. Mater. Sci.*, **44,** 1219-1236 (2009).

[20]S. Munoz and U. Anselmi-Tamburini, Temperature and Stress Fields Evolution during Plasma Sintering Processes, *J Mater Sci.*, **45,** 6528-6539 (2010).

[21]K. Matsugi, H. Kuramoto, O. Yanagisawa, and M. Kiritani, A Case Study for Production of Perfectly Sintered Complex Compacts in Rapid Consolidation by Spark Plasma Sintering, *Mater. Sci. Eng. A*, **354,** 234-242 (2003).

[22]V. Tikare, M. Braginsky, D. Buvard, and A. Vagnon, Numerical Simulation of Microstructural Evolution during Sintering at the Mesoscale in a 3D Powder Compact, *Comp. Mat. Sci.*, **48,** 317-325 (2010).

[23]J. Ziegler and N. Nichols, Optimum Settings for Automatic Controllers, *Trans. ASME*, **64,**759-768 (1942).

AN INTEGRATED VIRTUAL MATERIAL APPROACH FOR CERAMIC MATRIX COMPOSITES

G. Couégnat[a], W. Ros[a,b], T. Haurat[a], C. Germain[b], E. Martin[a], G.L. Vignoles[a]

[a] University of Bordeaux, LCTS, 3 allée de la Boétie, F-33600 Pessac, France
[b] University of Bordeaux, IMS, 350 Cours de la Libération, F-33410 Talence, France

ABSTRACT
　　We propose an integrated modeling software suite aiming at a numerical multi-scale simulation of ceramic matrix composites (CMC). This software suite features simulation capabilities for (i) reinforcement weaving design, (ii) matrix infiltration and (iii) mechanical and thermal behavior featuring damage. Numerical methods have been developed either on unstructured meshes or in voxel grid meshes (i.e. plain 3D images), or both. All simulations except matrix deposition by chemical vapor infiltration (CVI) have been developed in Finite Elements formulations; CVI simulation has been coded in random walk (RW) algorithms. All simulations are developed in strong relationship with the material microstructure in order to provide a sound basis for experimental validation with respect to actual material samples.

INTRODUCTION

　　Ceramic-matrix composites (CMC) have been shown to be promising candidates for civil jet engine hot parts[1], among other applications, thanks to their excellent lifetime in severe environment[2]. However, massive industrial production of such CMC parts will be economically viable only if the fabrication cost is controlled while achieving sufficient confidence in the material lifetime duration. Thus, the optimization of the material structure and fabrication process is a key issue for further development of CMC materials. Unfortunately, the design parameters that can be optimized, such as the preform weaving pattern, the number, thickness or composition of matrix layers, etc., are numerous and the production of material samples is very expensive and time consuming. Therefore, a material development cycle merely based on experimental characterization is not feasible and should be supported by a companion modeling approach. A numerical prediction of the variation of the material properties when changing some design parameter would be fast and inexpensive provided it is reliable enough.

　　The most crucial point to ensure the reliability of the numerical predictions is the accuracy in the description of (i) the material architecture, and (ii) of the physical phenomena (mechanical and chemical). To address these requirements, all simulations presented here are developed in strong relationship with image analysis and synthesis to provide a sound basis for experimental validation with respect to actual material samples. Simulations could be performed either directly on 3D representations of the material, such as those produced by ultra-sound or tomographic investigations[3], or using idealized (yet realistic) models of the material microstructure. In the latter case, scanned samples images are analyzed to determine appropriate morphological and statistical properties of an actual material that could be used to generate an equivalent model microstructure. Model microstructures allow a better a priori control on the complexity and the geometric properties (e.g. periodicity) of the generated numerical models. They are usually sufficient when estimating effective properties. Yet, direct simulations are of greater interest when one wishes to capture local effects that could be missed in an a priori approach. In addition to these elements, since the material architecture is organized upon several length scales, multi-scale analysis and change-of-scale methods have to be performed. The numerical tools have been developed either on structured or on unstructured meshes. Indeed, in some cases, unstructured meshes are of larger interest, for example, in the case of ideal structure design; while on the other hand, if one starts from a real 3D image of the material structure, structured meshes like regular 3D grids are of easier use. Since our computational tools are compatible with structured and unstructured meshes, we have developed converters capable of e.g. creating a voxel grid from an unstructured FE mesh. In this way, exchange between these two kinds of representations is made possible.

METHODS AND TOOLS

Figures 1 and 2 show how our modeling approach is structured respectively at the fiber and fabric scale; large similarities are present, though the material geometries and the details of the physics are distinct. One of the crucial points is the possibility to work either from acquired images, which allows validation of the computations, or with direct user input, which confers the software package the character of a material design toolbox.

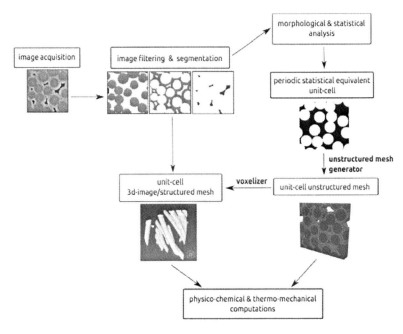

Figure 1. Developed virtual material approach at the fiber scale

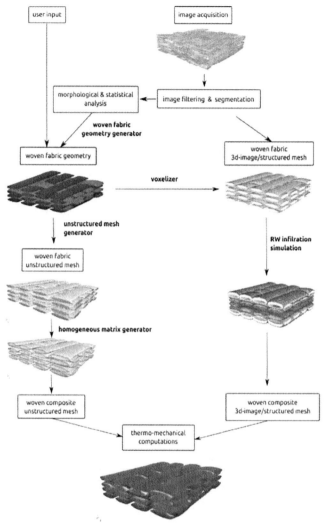

Figure 2. Developed virtual material approach at the fabric scale

Reinforcement Weaving Design

The fibrous reinforcement of CMC is usually made of a textile woven fabric of carbon or ceramic fibers. The weaving patterns range from simple 2D fabric (as twill or satin) to more complex 3D multilayered interlock weaves. A dedicated modeling package[4] has been developed allowing automatic generation of solid models for woven reinforcements (Figure 3, left). The internal geometry of the textile fabric is retrieved from the weaving pattern and basic geometric data about the tows (cross section shape, thickness, etc.). The tows interactions are deduced from the topology of the fabric, and the positions of their centerlines are then computed between each crossing points using an energy minimization procedure. The resulting geometric model can then be converted to either an unstructured or a structured mesh. The meshing process is straightforward since the underlying geometric model guarantees that no interpenetration occurs between the tows entities. Moreover, the local material properties are assigned automatically depending on the local crimp angle of the tows. In the current geometric approach, the tows are considered as rigid entities, i.e. not accounting for the experimentally observed changes of cross section[5].

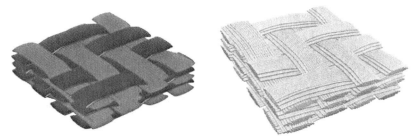

Figure 3. Numerical representation of a woven fabric obtained using the geometric model (left) and the intermediate fibrous model (right).

An on-going development consists in a more realistic representation of the tows as an intermediate fibrous model built upon the geometric one. The tows entities are decomposed as sets of individual "macro-fibers" and the initial fabric geometry is relaxed using a dynamic relaxation algorithm to simulate the tensile load applied to the fabric during the weaving process. Contact and friction forces between each individual "macro-fibers" are taken into account, which leads to more realistic model of the textile preform, including tows compaction, cross-section changes and fibers reorientations, as illustrated in Figure 3 (right).

Design of the Fiber Arrangement in the Tows

The arrangement of fibers in the tow, as opposed to the arrangement of the tows in the fabric, does not apparently follow a rigorous geometrical specification – rather, it appears more as a random process. So, in this work, the characteristics of this random arrangement have been extracted from the actual material by image analysis. In the case of carbon fibers, this is rather a difficult task, which has led to specific image processing developments[6,7]. Then, in a 2D perpendicular view of the tow, the fibers are identified, and their quasi-random arrangement is characterized using a two-point probability function as a statistical descriptor. Based on this descriptor, spatial correlation and length scales of the microstructure could be inferred. Statistical equivalent unit cells could then be generated using this information: starting from a random initial configuration, fibers positions are optimized so that the two-point probability function of the unit cell replicates the one of the initial microstructure, and that the fibers are eventually forbidden to overlap. This ensures that the reconstructed unit cells exhibit both the expected volume fraction and microstructure characteristic lengths. An automatic meshing procedure has also been developed to produce unstructured meshes of complete unit cells, i.e. including interphase material and one or several layer of matrix around the fibers[8].

Matrix Infiltration Simulation

Two levels of detail for the simulation of matrix infiltration by isothermal Chemical Vapor Infiltration (I-CVI) have been implemented. The first-level model consists in filling the virtual tows by a constant amount of matrix, and in covering them by a seal-coat layer with constant thickness (see Figure 2, left). For unstructured mesh representations of the woven fabric, this simulation is performed by Boolean operations directly on meshes[9].

Figure 4. Macro-scale (top) and micro-scale (bottom) infiltration modeling by random walkers.

A second level of simulation involves the actual physico-chemical modeling of chemical vapor infiltration. A two- scale approach has been developed, based on X-ray CMT images or on synthesized 3D images discretized on a regular voxel grid[10]. An intra-tow code allows for the precise simulation of the matrix deposition[11,12]; the deposit thickness variations are shown to depend on (i) the diffusion-reaction competition, which implies a smaller deposit thickness in small pores, and (ii) the pore network connectivity. The results are interpreted as local laws for effective parameters (diffusivity, bulk reactivity)[13] in a large-scale solver based on Brownian motion[14] capable of simultaneously computing the matrix phase amount in the tows and the seal-coat matrix thickness outside the tows. This code takes into account the local values of the volume fraction and of the orientation of the fibers. Both codes have been validated with respect to experimental data[11,14] in the case of Carbon-carbon composite fabrication; they are currently being applied to CMC, as illustrated on Figures 2 and 4.

Computation of Effective Mechanical Properties

The generated micro and meso cell meshes of a CMC material could be used to evaluate its initial and effective thermo-mechanical properties during its lifetime at both scales. Once the meshes are constructed and the local materials properties are assigned to the elements, the effective mechanical properties can be computed in a relatively straightforward manner using finite element calculation (see e.g. Figure 5, left). However, the large size of the resulting meshes motivates the use of advanced numerical tools like parallel solver to keep the computation time acceptable.

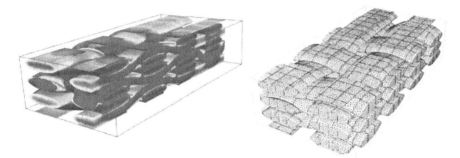

Figure 5. Stress distribution in a woven CMC material (left) and FE mesh of the same material with several crack families inserted (right).

When they are subjected to a thermo-mechanical loading, CMC materials usually experience a fragmentation process and exhibit progressive damage. To evaluate the effect of this damage development on the effective mechanical properties, cracks and debonding corresponding to damage states observed experimentally could be inserted directly into the FE meshes (see e.g. Figure 5, right). The evolution of the 3D elasticity tensor could then be retrieved as response surfaces for any damage state by interpolation. The numerical evaluation of the so-called damage effect tensor served as a basis for the development of a multi-scale damage model[15,16], where the crack densities and the debonding lengths are directly used as the damage variables. The effective material behavior model is written in a classical macroscopic framework, but it benefits from the multi-scale nature of the damage effect tensor and thus allows access to the evolution of the material microstructure at the large (fabric) scale.

Computation of Effective Thermal Properties
The main thermal property to be evaluated is the heat conductivity, or alternatively, the heat diffusivity, since the heat capacity may be easily computed and measured. This may be carried out from the knowledge of the individual properties of the components and of their arrangement by several averaging methods. Here, using the same method than for mechanical computations, the effective thermal conductivity of the undamaged material, and its evolution with damage development, can be straightforwardly estimated using finite elements[17].
The elementary mechanisms of heat transfer in CMC may be not only conductive, but also may include radiation at high temperatures. In this case, the effective heat conductivity of the material is enhanced by radiation through the material pore network. To address this, we are currently developing a solver based again on random walkers, allowing for a coupled conductive (i.e. diffusive)/radiative transfer. Simulated conduction in the solid accounts for the local anisotropy of the tows, and follows primarily the orientation of the fibers, which has been detected by classical image analysis tools. As a result, effective conductivities with and without radiation may be evaluated from the same image[18].

INTEGRATED MODELING APPROACH
We now present how the different modeling tools introduced above could be used in an integrated manner to produce a numerical model of a virtual CMC material.

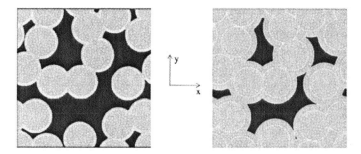

Figure 6. An example of a representative unit cell at fiber scale with different matrix thickness for the computation of evolution of the effective mechanical properties as a function of the intra-tow porosity.

- From high resolution tomographic images, or optical micrographs, of typical CMC tows, the local diffusive parameters are estimated using the intra-tow random-walk code (Figure 4, bottom). This allows to infer effective laws for the diffusivity and the reactivity as a function of the local intra-tow porosity.
- Using the same images, the statistical descriptors of the fiber distribution are evaluated and statistical equivalent unit cells are generated using this information (Figure 6). The effective mechanical properties are then evaluated for an increasing volume fraction of matrix, which allows to derive the evolution of the tow elastic properties as a function of the local intra-tow porosity.
- The geometry of a textile reinforcement is simulated using the woven fabric generator and converted into a voxel representation. For each voxel, the local fiber orientation and porosity are assigned automatically (Figure 7).
- The simulation of the matrix deposit is performed using the large-scale infiltration code. The local effective diffusive and reactive parameters are estimated from the volume fraction and fiber orientation in each voxel using the effective laws plotted in step (1).
- When the infiltration simulation is complete, effective elastic properties could be attributed to each voxel. For voxels which initially resided in tow material, the effective mechanical properties are evaluated by considering (i) the local porosity at the end of the simulation (given by the voxel grey level) using the functions identified in step (2) and (ii) the local fiber orientation.
- Finally, numerical tests could be performed to evaluate the effective mechanical or thermal properties of the virtual CMC material at the fabric scale (see e.g. Figure 8).

Figure 7. Voxel representation of a raw woven fabric (left) and corresponding fiber orientation (right)

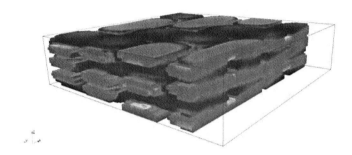

Figure 8. Stress distribution in a virtual CMC material for a tensile test. The large-scale residual porosity is 30% and the corresponding estimated Young's modulus is 140GPa which are typical values for this family of material.

CONCLUSION

We have introduced an integrated software suite and modeling approach aiming at a numerical multi-scale simulation of CMC material. It features (i) reinforcement weaving design, (ii) matrix infiltration simulation, (iii) and computation of effective mechanical and thermal properties, eventually including damage effect. All simulations are developed in strong relationship with image analysis and synthesis, in order to provide a sound basis for experimental validation with respect to actual material samples. This software suite can be used to predict the material behavior as a function of its structural parameters and of the properties of each individual phase, thus providing a first attempt at a "virtual material toolbox" environment. Of course, numerous improvements, tests, and validations have to be still performed, which are the object of ongoing and future work.

REFERENCES
[1] F. Christin, A global approach to fiber nD architectures and self-sealing matrices: from research to production, *Int. J. Appl. Ceram. Technol.*, **2**, 97–104 (2005).
[2] J.C. Cavalier, I. Berdoyes, and E. Bouillon, Composites in aerospace industry, *Adv. Sci. Technol.*, **50**, 153–162 (2006).

[3] J. Kim, P.K. Liaw, D.K. Hsu, and D.J. McGuire, Nondestructive evaluation of Nicalon/SiC composites by ultrasonics and X-ray computed tomography, *Ceram. Eng. Sci. Proc.*, **18**, 287-296 (1997).

[4] G. Couégnat, E. Martin, and J. Lamon., Multiscale modelling of the mechanical behaviour of woven composite materials, in *17th International Conference on Composite Materials, Edinburgh, UK*, paper ID11.5 (2009).

[5] P. Badel, E. Vidal-Salle, E. Maire, and P. Boisse, Simulation and tomography analysis of textile composite reinforcement deformation at the mesoscopic scale, *Compos. Sci. Technol.*, **68**, 2433–2440 (2008).

[6] C. Mulat, M. Donias, P. Baylou, G. L. Vignoles, and C. Germain, Optimal orientation estimators for detection of cylindrical objects, *Signal Image and Video Processing,* **2**, 51-58 (2008).

[7] C. Mulat, M. Donias, P. Baylou, G. L. Vignoles, and C. Germain, Axis detection of cylindrical objects in three-dimensional images, *J. of Electronic Imaging,* **17**, 031108-031108 (2008).

[8] G. Couégnat, E. Martin, and J. Lamon, 3D multiscale modeling of the mechanical behavior of woven composite materials, *Ceram. Eng. Sci. Procs.*, **31**, 185-194 (2010).

[9] G. Couégnat, Multiscale modeling of the mechanical behavior of woven composite materials (in French), Ph.D. thesis, Université Bordeaux 1 (2008).

[10] G. L., Vignoles, C. Mulat, C. Germain, O. Coindreau, and J. Lachaud,. Benefits of X-ray CMT for the modelling of C/C composites, *Adv. Eng. Mater.* **13**, 178–185 (2011).

[11] G. L., Vignoles, W. Ros, C. Mulat, O. Coindreau, and C. Germain, Pearson random walk algorithms for fiber-scale modeling of Chemical Vapor Infiltration, *Comput. Mater. Sci.* **50**, 1157-1168 (2011).

[12] G. L., Vignoles, C. Germain, O. Coindreau, C. Mulat, and W. Ros, Fibre-scale modelling of C/C processing by Chemical Vapour Infiltration using X-ray CMT images and random walkers, *ECS Transactions* **25**, 1275-1284 (2009).

[13] G. L., Vignoles, C. Germain, C. Mulat, O. Coindreau, and W. Ros, 2010. Modeling of infiltration of fiber preforms based on X-ray tomographic imaging, *Adv. Sci. Technol.*, **71**,108-117 (2010).

[14] G. L., Vignoles, W. Ros, I. Szelengowicz, and C. Germain, A Brownian motion algorithm for tow scale modeling of chemical vapor infiltration, *Comput. Mater. Sci.*, **50**, 1871-1878 (2011).

[15] G. Couégnat, E. Martin and J. Lamon, Multiscale modelling of woven ceramic matrix composites based on a discrete micromechanical damage description, in *4th European Conference on Computational Mechanics*, Paris, France, CDROM ref. 1358 (2010).

[16] P. Pineau, G. Couégnat, and J. Lamon, Virtual testing and simulation of multiple cracking in transverse tows of woven CMCs, *Ceram. Eng. Sci. Procs,* **31**, 319-328 (2010).

[17] J. El Yagoubi, J. Lamon, J.-C. Batsale, Multiscale Modelling of the Influence of Damage on the Thermal Properties of Ceramic Matrix Composites, *Adv. Sci. Technol.* **73**, 65-71 (2010).

[18] G. L. Vignoles, J.-F. Bonnenfant, I. Szelengowicz, and L. Gélébart, 2010. Prédiction de la diffusivité thermique à haute température d'un composite SiCf/SiC : un outil numérique basé sur des marches aléatoires hybrides, in *Matériaux 2010 Conference,* CDROM, 1526 (in French) (2010).

A NEW ANISOTROPIC CONSTITUTIVE MODEL FOR CERAMIC MATERIALS FAILURE

S Falco[a*], C E J Dancer[b], R I Todd[b], N Petrinic[a]

[a] Department of Engineering Science, University of Oxford, Parks Road, Oxford, OX1 3PJ, UK
[b] Department of Materials, University of Oxford, Parks Road, Oxford, OX1 3PH, UK

*Corresponding author: simone.falco@eng.ox.ac.uk

ABSTRACT

The lack of understanding of true deformation and failure mechanisms in ceramic materials for armour applications hinders the development of novel materials and the corresponding potential for improved armour solutions. Numerical modelling plays a key role in gaining the required understanding at all relevant length scales. While micro-scale modelling provides new insights into the role of various material characteristics, at the macro-scale such information enables more accurate simulation of the ballistic material behaviour when designing components and structures. Nevertheless, most of the existing macroscopic material models for the numerical simulation of ceramic armour are still based on continuum approaches assuming isotropic material response to impact loading. The constitutive model presented in this work allows for non-isotropic behaviour by proposing the damage as a tensor. Moreover the model incorporates aspects of the physics governing the compressive damage by basing the damage tensor evolution on considerations of crack growth at lower scales.

INTRODUCTION

Ceramic materials are currently used for both personal and vehicle armour protection since they can be very effective at stopping ballistic projectiles by breaking and/or eroding them[1]. However, ceramic armours are fairly heavy and lack multi-hit capability, mainly due to their excessive fragmentation during the impact. Limited understanding of the mechanism involved in ceramic failure at high strain rate prevents the determination of its ideal properties for armour applications, thus hindering the development of new ceramics for armour.

Numerical modelling is crucial to understand high strain rate performance of ceramic materials. Modelling plays a dual role: it helps design the experiments on a macroscopic scale and gives insights into the role of micro-mechanism in determining ballistic performance.

In the last years several constitutive models have been developed in order to simulate numerically the behaviour of ceramic material at high strain rates and their failure mechanisms[2]. The various models available can be categorized in two classes: those based on a continuum approach and those based on a discrete approach[3]. Even if the discrete models better simulate the crack behaviour they are computationally too expensive to be use in structural level investigations.

Within the continuum approach, in general, the damage can be "Phenomenological" or "Micro-mechanically motivated"[4]. In the first case the ceramic material is simulated as an elastic-plastic medium subject to a damage mechanism that reduces the strength as the deformation proceeds. The second class instead links the damage evolution with micro-mechanical considerations about the failure mechanism in the material (e.g. crack growth).

In this article we develop a constitutive model for ceramic materials that takes account of the crack growth mechanism involved in the failure of ceramic materials and of the anisotropy intrinsic in the phenomena.

CONSTITUTIVE MODEL

The anisotropic, micro-mechanically motivated model presented in this paper is an elaboration of one of the approaches developed by Johnson and Holmquist, particularly of the so-called "modified Johnson-Holmquist model[5]" (hereinafter referred to as "JH2 model").

93

The development process of the constitutive model for ceramic materials presented in this paper is articulated in two parts. First the capability of a non-isotropic evolution of the damage has been implemented. Then the damage evolution has been linked to micro-mechanical considerations on cracks evolution.

JH2 model
JH2 model defines the damage analogously to the Johnson-Cook[6] model for metals. It is a scalar value that increases proportionally to the plastic strain increment following the law:

$$D = \sum \frac{\Delta \varepsilon_p}{\varepsilon_p^f} \quad (1)$$

where $\Delta \varepsilon_p$ the incremental plastic strain and $\varepsilon_p^f = f(P)$ the plastic strain to fracture under constant pressure (function of the pressure). As the damage increases the yield surface shrinks from the intact condition ($D = 0$) to the completely failed condition ($D = 1$), as shown in Figure 1:

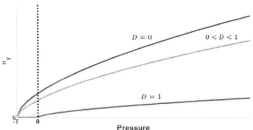

Figure 1: Strength definition in JH2 model

The yield stress (σ_Y) decreases as the damage increases with the following law:

$$\sigma_Y = \sigma_i - D(\sigma_i - \sigma_f) \quad (2)$$

σ_i is the strength of the intact material, while σ_f is the strength of the completely failed material – clearly only in compression. They are defined as:

$$\begin{cases} \sigma_i = A(P^* + T^*)^N \cdot \sigma_{HEL} \\ \sigma_f = B(P^*)^M \cdot \sigma_{HEL} \end{cases} \quad (3)$$

where σ_{HEL} is the stress at the Hugoniot Elastic Limit (HEL), P^* and T^* the pressure and the maximum tensile strength normalized on the pressure at HEL, A, B, M and N material constants.
If the trial equivalent stress (i.e. Von Mises stress calculated assuming that the material is elastic) overcomes the yield stress defined in (2), the trial stresses are radially returned on the yield surface and the damage increases because of the plastic strain increment as in (1).
The model does not embody either microstructure or normative material properties (e.g. toughness and hardness), since is based on multiple exponents and coefficients calibrated through dynamic measurements. Nevertheless JH2 model simulates with adequate fidelity many aspects of the impact on ceramic materials. In 2003 JH2 was implemented in the commercial FEM code LS-Dyna[7],

and nowadays it is the unique constitutive model for ceramic materials in the software, making it probably the most widely used in the industry.

The development of the new constitutive model started from the implementation of the JH2 model as a "User Defined Material model[8]" in LS-Dyna[9]. The user material model – coded in FORTRAN – has to provide the relationship between strains and stresses, the damage initiation criterion and the damage evolution law[10].

The user material model is validated against the results of simulations run using the JH2 model already implemented in LS-Dyna. The test cases shown (Figures 2 and 3) are two of those suggested by Johnson and Holmquist in the paper in which the JH2 model was first presented[4].

Both the test cases consist of a single cubic element, constrained on all the sides but the upper one, on which a pressure load is applied as shown in Figure 2. In "Test case B" the material does not have any strength after the complete failure, while in "Test case C" the material retains reduced strength in compression after the failure as explained previously in this section.

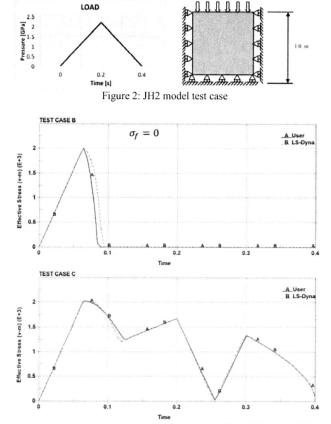

Figure 2: JH2 model test case

Figure 3: Comparison of test case results using the "User material" and LS-Dyna JH2 material model

As can be seen in Figure 3 the implemented material model (full line) behaves practically the same as the JH2 model already implemented in LS-Dyna (dashed line).

Adding anisotropy

One of the biggest limitations of the JH2 model is that the material is considered to be always isotropic, even if in reality the damage in ceramics is mainly due to propagation of cracks inside the material, which is an intrinsically anisotropic mechanism. For this reason the "user material" was modified to let the damage evolve differently in the different directions.

This goal was achieved by defining the damage increment as a diagonal tensor rather than as a scalar value, as presented in Eq (1).

$$dD = \begin{bmatrix} dD_x & 0 & 0 \\ & dD_y & 0 \\ & & dD_z \end{bmatrix} \quad (4)$$

The damage increment in the different directions ($j = x, y, z$) is defined as:

$$dD_j = \frac{\Delta\varepsilon_{pj}}{\varepsilon_f^p} \quad (5)$$

where

$$\Delta\varepsilon_{pj}^2 = \Delta\varepsilon_{p(j,x)}^2 + \Delta\varepsilon_{p(j,y)}^2 + \Delta\varepsilon_{p(j,z)}^2 \quad (6)$$

and is linked with the total damage by the relation:

$$dD = \sqrt{dD_x^2 + dD_y^2 + dD_z^2} \Rightarrow D^{(n+1)} = D^{(n)} + dD \quad (7)$$

After having defined the damage increment in three orthogonal directions, the yield stress in those directions (σ_{Y_j}) can be defined as:

$$\sigma_{Y_j} = \sigma_{i_j}^{(anis)} \cdot \sqrt{1-D} \cdot \sqrt{1 - \sum \frac{dD_j^2}{dD}} \quad (8)$$

The intact material strength in the three directions ($\sigma_{i_j}^{(anis)}$) needs to be calibrated. Preliminarily – assuming that the material is initially isotropic (i.e. $\sigma_{i_x}^{(anis)} = \sigma_{i_y}^{(anis)} = \sigma_{i_z}^{(anis)}$) – the intact strength of the material in any direction can be linked to the one defined in JH2 model (isotropic) through a calibration parameter β.

$$\sigma_i^{(iso)} = \sqrt{\beta} \cdot \sigma_i^{(anis)} \quad (9)$$

The effect of β on the material behaviour will be explored later in this section.

Similarly to the damage and the yield stress, the effective trial stress ($\bar{\sigma}_j^{tr}$) is defined along three orthogonal directions:

$$\bar{\sigma}_j^{tr\,2} = S_{(j,x)}^{tr}{}^2 + S_{(j,y)}^{tr}{}^2 + S_{(j,z)}^{tr}{}^2 \qquad (10)$$

Following the classic JH2 model approach, if the effective trial stress ($\bar{\sigma}_j^{tr}$) overcomes the yield stress (σ_{Y_j}), the deviatoric stresses are radially returned on the yield surface. The difference with the isotropic approach is that the stresses are returned separately along three orthogonal directions:

$$if\ \bar{\sigma}_j^{tr} > \sigma_{Y_j} \Rightarrow S_{jk} = \frac{\sigma_{Y_j}}{\bar{\sigma}_j^{tr}} S_{jk}^{tr}, \qquad with\ k = 1\dots3 \qquad (11)$$

In tension, instead, the material fails suddenly ($D = 1$) when the effective trial stress overcomes the yield stress.

Next pictures report simulations with isotropic and anisotropic constitutive models. In Figure 4 the test case is the same single cubic element of Figure 2, simply subject to a different load.

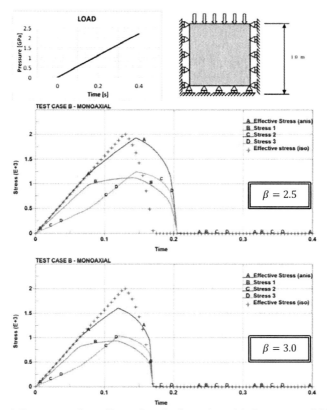

Figure 4: Stress comparison of isotropic and anisotropic models for a mono-axial load

In Figure 4 the "Stress 1" (i.e. the stress in the load direction) increases faster than the other two stresses (which are identical because of the symmetry) and is the first to reach the critical stress value. Once the material starts failing in one direction, the stress rate in the other two directions increases markedly due to the confinement. Lastly when "Stress 2" and "Stress 3" also reach the critical stress value, the effective stress drops and the material quickly fails.

For the same time of failure (i.e. $\beta = 3$) the maximum effective stress predicted from the anisotropic model is appreciably lower than the one predicted by the isotropic model because the failure is led by damage in the direction with the highest stress (i.e. load direction) and not by a scalar value representing the total damage in the solid. Moreover the failure process (i.e. from the first time $D \neq 0$ to $D = 1$) using the anisotropic model is slower because the presence of damage along one direction can coexist with the absence of damage along the other directions.

In Figure 5 the test case is the same single cubic element of Figure 2 and Figure 4 with the right side unconstrained. P_1 and P_2 are the loads applied respectively on the upper and right sides, and in this particular test they are perfectly equal.

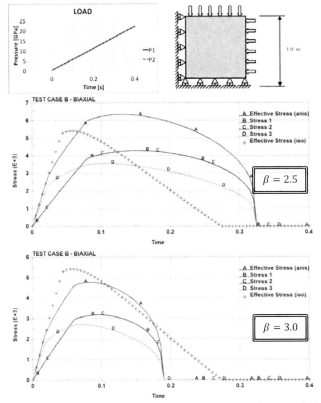

Figure 5: Stress comparison of isotropic and anisotropic models for a bi-axial load

In Figure 5, instead, the "Stress 3" grows faster than the stress in the load directions – due to the effect of the confinement – and reaches first the critical stress. Subsequently the critical stress is reached by "Stress 1" and "Stress 2" too. Those behaviours are reflected in the effective stress curve, first as a reduction of the stress rate (i.e. damage initiation in "Stress 3" direction), then in the drop of the curve until the complete failure of the material.

As for the mono-axial load test case the failure process using the anisotropic model is slower than the in the isotropic case.

Moreover in all the figures it is evident that the model has a big sensitivity to β, therefore some tests are required to calibrate the intact material strengths $\sigma_{i_j}^{(anis)}$.

The capability of the model for complex simulations was tested by carrying out a Hertzian test simulation (Figure 6).

Figure 6: Equivalent stress during Hertzian indentation

Finally, it should be noted that with the proper calibrations the model also has the capability to include any initial anisotropy of the material (e.g. due to the manufacturing process) by defining different $\sigma_{i_j}^{(anis)}$ in the three orthogonal directions.

New "crack-based" damage evolution

The last improvement to the model presented in this paper links the damage evolution with micro-mechanical considerations relating to the growth of cracks.

Ceramic materials contain defects (e.g. micro-cracks, small holes, poorly bonded particles) that can act as the nuclei for new cracks when the material is loaded. While the range of possible nuclei is quite wide[11] their behaviour in compression can be well described by isolated sharp inclined cracks[12,3] (Figure 7).

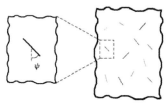

Figure 7: A population of cracks in a solid subject to compression loads, growing in the maximum principal stress direction[10]

Let's consider the sample volume subject to a purely compressive load (i.e. all the principal stresses are compressive), with σ_1 the most compressive principal stress, and with:

$$|\sigma_1| > |\sigma_2| > |\sigma_3|$$

The remote load creates a stress state on the crack that can be composed in shear stress (τ) and normal stress across the crack (σ) (Figure 8a), defined as:

$$\tau = (\sigma_3 - \sigma_1) \sin \psi \cos \psi$$
$$\sigma = \sigma_1 \sin^2 \psi + \sigma_3 \cos^2 \psi$$

The stress state results in an opening force F_W acting in direction of σ_3 (Figure 8b):

$$F_W = (\tau + \mu\sigma)\pi a^2 \sin \psi$$

where μ is the friction coefficient and a the crack radius.

F_W creates a mode I stress intensity inducing the crack to grow in direction of σ_1 (Figure 8c).

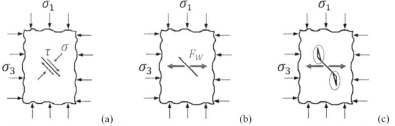

(a)   (b)   (c)

Figure 8: Stress state on the crack (a), opening force created by the stress state (b) and crack growing direction induced by the opening force (c)

Embedding these considerations in the anisotropic damage evolution approach defined in the previous section, the damage can be assumed to evolve only in the direction of the most compressive principal stress (X_1). Since the principal directions can rotate with respect to both global and local directions because of the change of the remote load and the damage evolution, the model has to be developed such that it is able to follow these rotations.

The two main problems of evaluating the damage along the principal directions are: (i) tensor rotation is computationally expensive; (ii) each rotation has a small error, so rotating back and forth at each time step leads to an error accumulation that can affect the accuracy of the simulation.

The first issue was avoided by using a non-iterative method for the rotation of 3x3 symmetric matrices[12] which is less computationally expensive than the classic iterative matrix rotation method. In this method, the negated characteristic equation

$$0 = -\det(A - \lambda I) = -det \begin{bmatrix} a_{00} - \lambda & a_{01} & a_{02} \\ a_{01} & a_{22} - \lambda & a_{12} \\ a_{02} & a_{12} & a_{22} - \lambda \end{bmatrix} = \lambda^3 - c_2\lambda^2 + c_1\lambda - c_0 \qquad (12)$$

where

$$c_0 = a_{00}a_{11}a_{22} + 2a_{01}a_{02}a_{12} - a_{00}a_{12}^2 - a_{11}a_{02}^2 - a_{22}a_{01}^2$$
$$c_1 = a_{00}a_{11} + a_{00}a_{22} + a_{11}a_{22} - a_{01}^2 - a_{02}^2 - a_{12}^2 \quad (13)$$
$$c_2 = a_{00} + a_{11} + a_{22}$$

is solved by a change of variables. To eliminate the square term λ is substituted with:

$$\lambda = \xi + \frac{c_2}{3} \quad (14)$$

This leads to the equation:

$$\xi^3 + a\xi + b = 0 \quad (15)$$

where

$$a = \frac{3c_1 - c_2^2}{3}, \qquad b = \frac{-2c_2^3 + 9c_1c_2 - 27c_0}{27}, \qquad q = \frac{b^2}{4} + \frac{a^3}{27} \quad (16)$$

Finally, the generic roots of the equation are:

$$\lambda_0 = \frac{c_2}{3} + 2\sqrt{-\frac{a}{3}}\cos(\vartheta)$$
$$\lambda_1 = \frac{c_2}{3} - \sqrt{-\frac{a}{3}}\left(\cos(\vartheta) + \sqrt{3}\sin(\vartheta)\right) \quad (17)$$
$$\lambda_2 = \frac{c_2}{3} - \sqrt{-\frac{a}{3}}\left(\cos(\vartheta) - \sqrt{3}\sin(\vartheta)\right)$$

where

$$3\vartheta = atan2\left(\sqrt{-q}, \frac{b}{2}\right) \quad (18)$$

To further decrease the computational time the model provides simplified equations to evaluate the roots if $q = 0$.

The non-iterative method is comparable in accuracy to the iterative method and is about 4.5 times faster, which is a considerable saving in computational time[13].

The second issue of evaluating the damage along the principal directions is to avoid the small and negligible error of each rotation accumulating, leading to a significant error in simulations with a high number of time steps. Rotating from principal to local coordinate system only the stress radial return and the damage increment – updating stress and damage tensors in local coordinates – achieves the goal by keeping the magnitude of the quantity rotated low and avoiding back and forth rotation of any tensor.

Similarly to the method used in the classic JH2 model, the trial stress tensor σ_{ij}^{tr} is calculated from the strain tensor. Then the principal stresses (i.e. eigenvalues of the stress tensor) are evaluated as in Eq. 17 and exploited to calculate the eigenvectors $(Eig1, Eig2, Eig3)$ associated respectively with the maximum, the second and the minimum principal stress.

The rotation matrix ($[M]$) is:

$$[M] = [Eig1 \quad Eig2 \quad Eig3] \qquad (19)$$

The rotation matrix ($[M]$) is used to rotate the yield stress tensor in principal coordinates ($\sigma_Y^{(pri)}$).

$$\sigma_{Y_{ij}}^{(loc)} = \sigma_i(1 - D_{ij}) \Rightarrow \left[\sigma_Y^{(pri)}\right] = [M]^{-1}\left[\sigma_Y^{(loc)}\right][M] \qquad (20)$$

If along the maximum principal stress direction (X_1), the stress overcomes the yield stress, the stresses are radially returned on the yield surface. The returning happens in local coordinates by rotating the difference between the principal and the yield stresses ($d\sigma_1$) along X_1 from principal to local coordinates.

$$if\ \sigma_1 > \sigma_{Y1}^{(pri)} \Rightarrow d\sigma_1 = \sigma_1 - \sigma_{Y1}^{(pri)} \Rightarrow d\sigma_{ij} = [M]\begin{bmatrix} d\sigma_1 & 0 & 0 \\ 0 & 0 & 0 \\ 0 & 0 & 0 \end{bmatrix}[M]^{-1} \Rightarrow \sigma_{ij} = \sigma_{ij}^{tr} - d\sigma_{ij} \qquad (21)$$

Similarly the damage increment is evaluated in principal coordinates, rotated and added to the damage tensor in local coordinates.

$$dD_{ij} = [M]\begin{bmatrix} dD_1 & 0 & 0 \\ 0 & 0 & 0 \\ 0 & 0 & 0 \end{bmatrix}[M]^{-1} \Rightarrow D_{ij} = D_{ij}^{(old)} + dD_{ij} \qquad (22)$$

Finally – as for the anisotropic model presented in the previous section – the total damage is calculated as:

$$D = \sqrt{\sum(D_{ij}^2)} \qquad (23)$$

When $D = 1$ the material is considered completely failed.

By defining the intact strength (σ_i) as a tensor (with the proper calibration) instead of as a scalar the initial anisotropy of the material can be embedded into the model.

CONCLUSIONS

A constitutive model for characterizing the failure of ceramic materials is presented that considers the anisotropy of the damage evolution and the micro-mechanical behaviour of the material, as an improvement of the JH2 model.

First a method was presented to simulate the anisotropy of the damage evolution phenomenon by defining the damage as a tensor instead of as a scalar, and rewriting the equation to consider the damage evolving separately in different directions. The material model is compared with the JH2 model showing similar global predictions about the material behaviour, but highlighting the different failure mechanisms in three orthogonal directions.

Afterward some micro-mechanical considerations on crack growth from defects inside the material are presented, and included in the material model. The new material model involves the calculation of principal directions and matrix rotation, which are computationally expensive operations. To reduce the computational cost a non-iterative method to calculate eigenvalues and eigenvectors of a 3x3 symmetric matrix is embedded in the algorithm, and the model is modified to minimize the error deriving from rotations from principal to local coordinate systems.

ACKNOWLEDGEMENTS
This research was carried out as part of the Understanding and Improving Ceramic Armour (UNICAM) Project. We gratefully acknowledge funding from EPSRC and the Ministry of Defence, UK.

REFERENCES
[1] Dancer CEJ, Curtis HM, Bennett SM, Petrinic N, Todd RI (High strain rate indentation-induced deformation in alumina ceramics measured by Cr3+ fluorescence mapping) 2011: J. Euro Ceram Soc 31, 2177–2187
[2] Rajendran AM (Historical perspective on ceramic materials damage models) 2001: Proc. Ceramic Armor by Design Symposium, 281-297
[3] Espinosa H, Zavattieri PD, Dwivedi SK (A finite deformation continuum/discrete model for the description of fragmentation and damage in brittle materials) 1998: J. Mech. Phis. Solids 46, 1909-1942
[4] Deshpande VS, Evans AG (Inelastic deformation and energy dissipation in ceramics: A mechanism based constitutive model) 2008: J. Mech. Phis. Solids 56, 3077-3100
[5] Johnson GR, Holmquist TJ (An improved constitutive model for brittle materials) 1994: Proc. AIP Conference 309, pp. 981-984;
[6] Johnson GR, Cook WH (Fracture characteristic of three metals subjected to various strains, strain rates, temperatures and pressures) 1985: Eng. Fract. Mech. 21, 31-48
[7] Cronin D S (Implementation and validation of the Johnson-Holmquist ceramic material mode) 2003: Proc. 4th European LS-DYNA User Conference (DYNAmore)
[8] (LS-Dyna Keyword User's Material. 2007): Livermore Software Technology Corporation.
[9] Hallquist JO (LS-Dyna Theoretical Manual) 2006: Livermore Software Technology Corporation
[10] Gazonas G.A (Implementation of the Johnson-Holmqist II (JH-2) constitutive model into DYNA3D) 2002: Army Research Laboratory Report. ARL-TR-2699
[11] Evans AG, Meyer ME, Ferting KW (Probabilistic models for defect initiated fracture in ceramics) 1980: J. Nondestr. Eval. 1,111-122
[12] Ashby MF, Sammis CG (The damage mechanics of brittle solids in compression) 1990: PAGEOPH 133, 439-521
[13] Eberly D (Eigensystems for 3x3 Symmetric Matrices (Revisited)) 2011: Geometric Tools, LLC

STUDIES OF GAS-PHASE REACTIVITY DURING CHEMICAL VAPOR DEPOSITION OF BORON CARBIDE

G. Reinisch[1,2], J.-M. Leyssale[2], S. Patel[1], G. Chollon[2], N. Bertrand[1], C. Descamps[3], R. Méreau[4], G. L. Vignoles[1]

1. University Bordeaux 1
LCTS – Lab. for ThermoStructural Composites
3, Allée La Boëtie - F33600 PESSAC, France

2. CNRS
LCTS – Lab. for ThermoStructural Composites
3, Allée La Boëtie - F33600 PESSAC, France

3. Safran-Snecma Propulsion Solide
LCTS – Lab. for ThermoStructural Composites
3, Allée La Boëtie - F33600 PESSAC, France

4. CNRS
ISM – Institute of Molecular Sciences
316, Avenue de la Libération - F33410 TALENCE, France

ABSTRACT - Boron carbide is a key constituent of self-healing ceramic-matrix composites, due to its ability to fill matrix cracks by a liquid oxide phase upon oxidation. Its production for protective layers in a multilayered matrix can be performed by Chemical Vapor Infiltration. Since this process is difficult to manage in terms of deposit thickness and quality as a function of processing parameters, a modeling study has been developed. It features several steps: (*i*) *ab-initio* quantum chemical computation using the G3B3 method, (*ii*) derivation of thermochemical and kinetic data for the main species and reactions involved in the gas-phase mechanism, (*iii*) insertion of the obtained mechanism in a 0D thermochemical equilibrium solver and in a 1D chemical kinetic solver.

Several specific tools have been developed and applied for a correct treatment of step (*ii*). An extension of the hindered rotor models with variable kinetic function has been designed for a correct computation of partition functions associated to internal rotations in molecules or transition states. The B/3Cl/H subsystem has received special attention because of transition states without local energy maxima, internal rotations, and reaction product valley bifurcations.

Results are compared to experimental data and allow an identification of key species and main reaction pathways during deposition.

INTRODUCTION

Boron carbide draws interest since decades for several reasons[1]. Extremely hard, capable of neutron absorption, it has applications in armour[2] and nuclear[3] technologies. It is also employed as one type of the ceramic matrix layers in multilayer Self-Healing Ceramic-Matrix Composites (SH-CMCs) because of its ability to yield a crack-healing liquid oxide in a hot oxidizing atmosphere[4,5]. In the context of the latter application, boron carbide is often prepared by Chemical Vapor Infiltration (CVI), a variant of Chemical Vapor Deposition (CVD)[6]. As boron hydrides like BH3 are extremely toxic and hazardous, the most employed gaseous precursors are BBr3 [7] and mostly BCl3, mixed with hydrocarbons and diluted in hydrogen. CVD allows excellent material properties, but the control of the deposit thickness, stoichiometry and nanotexture requires a sound knowledge of the process. In this goal, numerous studies focused attention on the $BCl_3/CH_4/H_2$ system and on the boron carbide deposition from the latter.

Several experimental works [8-15] report apparent deposition kinetics. Karavan et al.[10,11] state that the principal product of the initial BCl_3 decomposition is $BHCl_2$ under atmospheric pressure; more recently, Berjonneau et al.[12-15] provided simultaneous characterizations of gas-phase composition, deposition kinetics and deposit nanotexture and composition, confirming and enriching the previous results. However, the complexity of the system response to variations of the experimental parameters prevents any extrapolation of the gas-phase behavior to unexplored conditions. Thus there is a large interest towards developing a simultaneous modeling approach.

Modeling works have dealt with thermochemical equilibrium computations [8,9,16-19] and on gas-phase reaction kinetics[20-23]. Exhaustive thermochemical computations[18,19] have confirmed $BHCl_2$ as a major gas-phase product and drawn the attention on the potentially important role of MethyDichloroBorane (MDB) in the gas-phase decomposition mechanism. Harris et al. [20] proposed a 31-reaction radical-based mechanism; in another context (B-doped Si deposition), Lengyel[21] has quantified the rates of three molecular reactions starting from BCl_3. After inspection of the literature, we have found that, though data on many species and reactions already exists, building a comprehensive of the gas-phase reactivity in the B/C/Cl/H system was still to be done. This paper will summarize our work done in this direction.

The first part will briefly present the overall strategy; the next parts deal with some details of the approach; finally, results are shown and discussed with respect to experimental data acquired at our lab.

MODELING STRATEGY

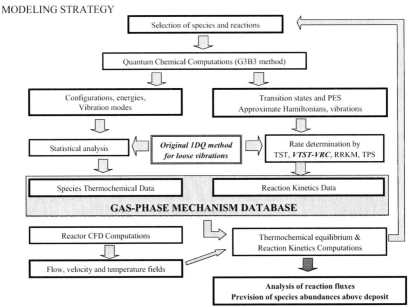

Figure 1. Flowchart of the modeling approach.

The global flowchart of the computational approach is depicted in Fig. 1. The final goal of the approach being the determination of the critical species responsible for the deposit quality and deposition rate, we require the achievement of thermochemical and reaction kinetics computations, possibly inserted in a CFD reactor model. This is based on the existence of a sound, consistent database describing the properties of all reactants and rate laws for all reactions. This in turn requires, in addition to a compilation of literature data (of experimental or theoretical origin), on the production of thermochemistry and reaction rate data. In the case of the B/C/Cl/H system, even though several published works are available, many reactions were missing; also, the evaluation of the data was sometimes requiring improvements due to some particularities, as described in the following part.

BUILDING THE THERMOKINETIC DATABASE

The starting point is the production of a consistent database for thermochemical data of the species (enthalpies, entropies, heat capacities), and for reaction rates. The first step is naturally the choice of the species taken into account. A rationale for our choice is to describe adequately the conditions met in CVI apparatuses, $i.e.$ moderate pressures (1-20 kPa) and temperatures (600-1300 K), and boron chloride, dihydrogen, and hydrocarbons as reactants. A critical examination of the large dataset studied by T. Wang $et\ al.$[19] led to the selection of a subset of species containing 1 boron and/or 1 or 2 carbons. This set includes H_2, HCl, Cl_2, BH_iCl_j ($0 \leq i+j \leq 4$) species, and some $BCCl_iH_j$ ($4 \leq i+j \leq 5$). Some diboranes and chlorodiboranes , as well as clusters with 9 or 12 central atoms (B or C) and H or Cl peripheral atoms have also been incorporated at first, but equilibrium computations have shown negligible amounts as soon as the Cl/B atom ratio was equal or superior to 3 [24]. Examination of the major species present as computed in thermochemical equilibrium computations has allowed the construction of a list of possible reactions, which have been split between molecular and radical reactions.

Feeding the database with consistent data rests on quantum chemical computations. First, equilibrium geometries, harmonic vibration frequencies and potential energy surfaces of a large set of chemical species in the system B/C/Cl/H are computed under the Density Functional Theory (DFT) framework at the B3LYP/6-31G(d,p) level of theory using the Gaussian 2003 package[25].

Vibration frequencies, known for being overestimated by DFT calculations, are scaled by the usual factor of 0.96, as confirmed by data from CCCBDB* on chloroboranes. Partition functions, entropies and heat capacities are obtained from standard statistical equations[26].

Central to the prediction of reaction rate constants is the calculation of the activation energies. Indeed, a variation of a few kcal/mol in the potential energy barrier between reactants and transition state can change a reaction rate by an order of magnitude. Calculation of energies thus requires numerical methods of high accuracy. In this work, we use the G3B3 composite method[27], in which the energy is obtained from single point calculations (at the equilibrium B3LYP/6-31G(d) geometry) at four different levels of theory (QCISD(T,E4T)/6-31G(d), MP4/6-31+G(d), MP4/6-31G(2df,p), MP2=Full/GTLarge), taking into account the electronic correlation to a large extent and a spin-orbit correction. Calculation with this method of $\Delta_f H°(298K)$ of 155 molecules has given a root mean square deviation of only 1.37 kcal/mol[28].

The contribution of some loose vibration modes (rocking and wagging) to the partition functions of several molecules has been computed, when necessary, with the 1DQ hindered rotor model[29]. This extension of previous hindered rotor (HR) models[30-32] is based on the use of a variable kinetic function in the Hamiltonian instead of a constant reduced moment of inertia. The variable kinetic function is first introduced in the framework of a classical 1D HR model, named 1DC. Then, an

---

* CCCBDB : Computational Chemistry Comparison and Benchmark Database, http://srdata.nist.gov/cccbdb

effective temperature-dependent constant is proposed in the cases of a quantum 1D (named 1DQ) and a classical 2D model (named 2DC). It has been obtained from a Maxwell– Boltzmann averaging of the real kinetic function. These methods allow a straightforward evaluation of vibration data (i.e. Density Of States of the vibration modes) of molecules, and of rate constants of reactions in which the transition state exhibits loose vibrations, without detailed scans of the multi-dimensional PES associated to these modes.

An example result is given for B_2Cl_4, one of the species in the CVD of boron carbide[18]. One of its vibration modes is a bending mode that the Harmonic Oscillator Approximation does not render satisfactorily at high temperatures. Fig. 2 displays the computation of the partial contribution of this bending mode to the total entropy of the molecule, as a function of temperature. Three models are compared: (i) free rotation, (ii) hindered rotation by the 1DQ method with a sinusoidal potential energy function, and (iii) harmonic oscillator. The analytical[33] and numerical computations of the reduced moment of inertia with 1DQ are in excellent agreement (I_{num}=289.1 amu.Bohr^2; I_{ana}=288.1 amu.Bohr^2). The torsional symmetry number of the rotation is 2 and the energy barrier is estimated as 1.7kcal/mol in order to match the torsional force constant at the equilibrium geometry obtained by normal mode analysis. As can be seen, the 1DQ method matches the Harmonic Oscillator model at very low temperatures and the free rotation model at very high temperatures.

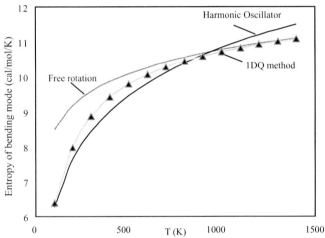

Figure 2 : Entropy of the bending mode in B_2Cl_4. Grey continuous line : free rotation. Black continous line : Harmonic Oscillator. Triangles : 1DQ Hindered Rotor model.

For reactions of the B/Cl/H submechanism, getting rate constants from quantum chemical computation has implied addressing several difficulties. Indeed, in several cases, there is no enthalpy maximum along the reaction coordinate, while there is a Gibbs function maximum. The Variational Transition State Theory (VTST)[34] has been used instead of the usual Transition State Theory (TST). The VTST approach consists in finding the Transition State (TS) by a minimization of the Gibbs energy of the intermediate state with respect to the reaction coordinate. The reaction coordinate is calculated using Variable Reaction Coordinate (VRC) framework[35,36], by optimizing the pivot points of a spherical dividing surface in order to minimize the associated Gibbs energy.

In our case, the determination of the orthogonal subspace of the trajectory points was rendered somewhat difficult by the fact that the involved vibration modes are loose and display a strong relative variation of their frequency. So, these loose vibrations in the transition states were treated by our *ad-hoc* 1DQ Hindered Rotor method.[29] Figure 3 illustrates some of the obtained results for the BCl_3 + H reaction system, which has concentrated all difficulties encountered in this study. [37] We see a potential energy hypersurface at 13 kcal/mol above $BHCl_3$, defined in the coordinate space given by the B-H distance, the B-Cl* distance, and the H-B-Cl* angle, where Cl* indicates the possibly leaving Cl atom. The Minimal Energy Path is illustrated by the dashed line; along it, a transition state is found (here, at 800 K: its position depends on temperature), and its associated dividing surface is represented: we can see that it is neatly curved, because the pivot point is not exactly located on the boron atom but lies inside the boron *p* orbital of $BHCl_2$.

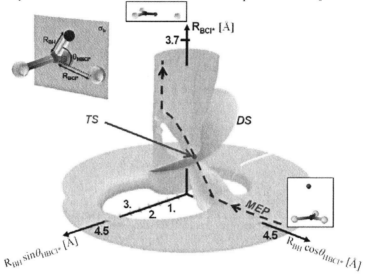

Figure 3: Potential Energy Hypersurface at 13 kcal/mol for the BCl_3 + H reaction using B3LYP/6-31G(d,p) level of theory. MEP : Minimum Energy Path, TS : Transition State, found by VTST-VRC method ; DS : Dividing Surface at 800 K, defining the orthogonal vibration modes (only the positive half is represented). Origin of energy taken at $BHCl_3$.

An additional result is that at higher energies the exit tube (centered on the RBCl* axis) bifurcates, leading to the possible evolution of two distinct sets of products from the same transition state, namely, (i) $BHCl_2$ + Cl, and (ii) → BCl_2 + HCl. Moreover, at lower energies, a poorly stable sp3 transient complex $BHCl_3$ may appear; its formation rate has been computed using the RRKM approach. Comparing it with the reported experimental rate of BCl_3 consumption through reaction with the H radical[38] allowed concluding that the correct activation energy has been found, while the tunneling effect is responsible for the large value of the pre-exponential coefficient at such a low pressure (2 torr). Dividing the tunneling frequency by a factor 2 allowed an excellent fit of the computed rate with respect to experiments.

These results are summarized in Figure 4, together with the Direct Abstraction (DA) reaction[21] $BCl_3 + H \rightarrow BCl_2 + HCl$, which was the only one previously reported. As can be seen in this figure, the first reaction to occur at low temperatures is not the DA reaction but is, in fact, the H addition reaction on BCl_3 leading to $BHCl_3$. At typical temperatures of CVD processes, from 800 to 1500 K, our study shows that BCl_3 is decomposed through the AE mechanism for which we have determined rate estimates for the first time. Since two sets of elimination products, namely $BHCl_2$ + Cl and BCl_2 + HCl, sharing the same TS, have been evidenced, a dividing coefficient has been computed using two different approaches, a statistical analysis of the PES and a Transition Path Sampling (TPS) method,[39,40] to obtain individual elimination rates from the total elimination rate.

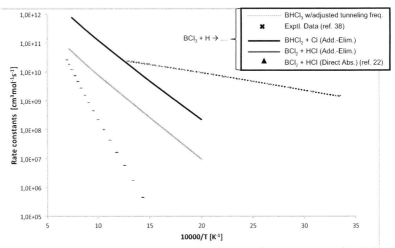

Figure 4. Summary of several computed and experimental rates of reactions starting from BCl_3 + H.

The same study has been carried out on the similar systems $BHCl_2$ + H and BH_2Cl + H. Bimolecular reactions, already studied previously[20], have been reassessed at the same level of theory as the other ones for consistency. The next step was to incorporate several reactions to account for the hydrocarbon chemistry. Species and reactions were taken from Norinaga et al..[42] Finally, reactions involving MethylDichloroBorane (MDB, CH_3BCl_2), the first B- and C-containing species susceptible to appear, have been evaluated and their rates added to the database. The result is a thermochemical and kinetic reaction mechanism[41] which can be used in various kinds of simulations. It features 19 species and 64 reactions.

GAS-PHASE EQUILIBRIUM AND REACTOR KINETICS COMPUTATIONS

All computations have been carried out using the Cantera freeware.[43] First, the results of 0D thermochemical equilibrium computations have helped confirming the choices made for the main species and reactions selection. Figure 5 displays as a function of temperature the computed mole fractions of the main species of the B/C/Cl/H gas-phase system at thermochemical equilibrium at P=12kPa and a B/C/3Cl/12H atom ratio, conditions representative of the CVD inlet mixture

$BCl_3+CH_4+4H_2$. At T<600K the gas is mostly BCl_3, CH_4 and H_2 (i.e. the intact precursors). When temperature further increases, BCl_3 and H_2 are consumed; they yield $BHCl_2$ and HCl, and in lesser amounts BH_2Cl then BH_3. It is clear that $BHCl_2$ and HCl are in similar amounts, HCl being slightly more abundant at the highest temperatures because of BH_2Cl production. The other boron-containing species are monoboranes and carboranes; all diboranes, BH and B are not represented since their mole fractions are below 10^{-8}. The hydrocarbons follow a classical route starting from methane, yielding ethane, ethylene, and acetylene in sequence. Thus, the carbon and the boron submechanisms are rather indifferent to each other; nonetheless, there is some carborane formation, principally MDB. There is quite no species containing a carbon-chlorine bond, since boron has a much larger affinity for chlorine than carbon. These results encourage discarding molecules containing more than one boron atom in the model while paying a special attention to MDB.

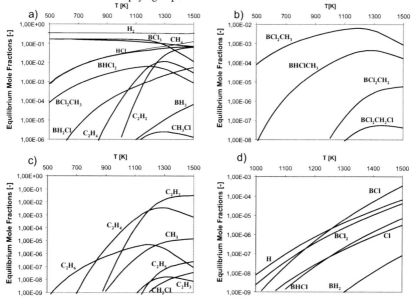

Figure 5. Computed homogeneous equilibrium mole fractions of gaseous species from a B/C/Cl/H mixture [B:C:Cl:H] = [1:1:3:12], at P=12kPa, and temperatures between 500 and 1500 K. a) Most abundant species; b) carboranes; c) less abundant hydrocarbons, d) H, Cl, and less abundant boron-containing species.

The next step was a 1D computation of the concentration profiles in the tubular reactor, using an experimentally determined thermal profile. The outlet values of the concentrations, at several temperatures, may be compared to experimental FTIR analyses obtained either in the case of BCl_3+H_2 mixtures[44] or of $BCl_3+CH_4+H_2$ mixtures.[45] Though diffusion may play a non-negligible role in practice (since the gas velocity is moderate), the results of 0D computations always were in close agreement with the 1D computations, provided that the 0D evolution time is adjusted to the 1D hot-zone residence time. Figure 6 collects typical results of this work. The hot-zone decomposition of the precursor mixture involves molecular reactions at low temperatures and moves on to radical reactions at higher temperatures. We can see on figs. 6a and b the radicals trapped in the hot zone. While a very reasonable agreement is obtained for the temperature evolution of the outlet concentrations of BCl_3 and $BHCl_2$, the model predicts the onset of non-negligible amounts of MDB at T > 1100K. This has encouraged an attempt to identify experimentally this molecule, which had been discussed in this context on a theoretical basis only.[19,20] The vibration frequencies were computed and the experimental FTIR spectra were searched for specific peaks. Two peaks at 833 and 1025 cm[-1] were found in close agreement with the computed frequencies (1036 and 830 cm[-1]). Figure 6d reports their relative evolution (in scaled units) as compared to the computed evolution, showing an excellent agreement until ~1200K A mismatch is observed above this temperature due to the starting deposition, not taken into account by this purely gas-phase model.. This gives an extra validation to the model. Finally, the computational results were analyzed in terms of reaction fluxes, as summarized in Figure 7 at 1450K in the hot zone, where the radical-based mechanism competes strongly with molecular reactions. It is seen that MDB formation arises simultaneously from the attack of CH_4 by BCl_3 and $BHCl_2$ (70%/30% proportions), the latter being formed jointly by the attack of H and H_2 on BCl_3.

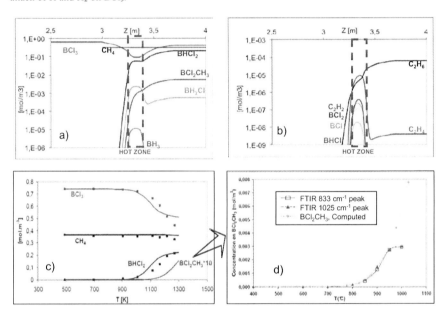

Figure 6: Reactor kinetics computations and validation. Conditions are as in figure 5. a) Concentration profiles of the main species along the tubular reactor position at T=1200K. The hot zone position is indicated by the shaded area. b) Concentration profiles of less abundant species. c) Measured (symbols) and computed (lines) outlet concentrations of major species. d) Comparison of the temperature evolutions of computed BCl_2CH_3 outlet concentration and of measured FTIR peak intensities at 833 cm[-1] and 1025 cm[-1].

CONCLUSION AND OUTLOOK

This paper has given an overview of an exhaustive modeling approach of the gas-phase reactivity in the B/C/Cl/H system involved in boron carbide chemical vapor deposition. Computations were performed at the quantum chemical, reaction dynamics, and macroscopic (thermochemical equilibrium and 0D/1D reactive flow kinetics) levels. A consistent thermochemical and kinetic database has been built; the mechanism has been validated against experimental data acquired by FTIR in a tubular hot-wall CVD reactor. The natural continuation of this work is to identify the deposition kinetics by comparing the gas-phase composition and the experimental mass gain and thickness evolution kinetics, and move to other systems, like Si/B/C/Cl/H for silicon/boron carbide deposition.[14]

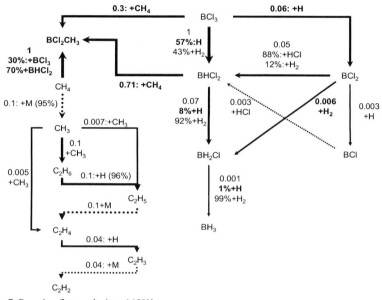

Figure 7. Reaction flux analysis at 1450K.

ACKNOWLEDGEMENTS

Snecma Propulsion Solide (Safran group) and CNRS are acknowledged for financial support through a PhD grant to G. R..

REFERENCES

[1] F. Thévenot, Boron Carbide—A Comprehensive Review, *J. Eur. Ceram. Soc.* **6**, 205-25 (1990).
[2] P.G. Karandikar, G. Evans, S. Wong, M. K. Aghajanian, M. Sennett, A Review of Ceramics For Armor Applications, *Ceram. Eng. Sci. Procs.* **29**, 159-74 (2009).
[3] K. Devan, A. Riyas, M. Alagan, P. Mohanakrishnan, A new physics design of control safety rods for prototype fast breeder reactor, *Ann. Nucl. Energy* **35**, 1484-91 (2008).

[4]R.Naslain, A.Guette, F.Rebillat, R.Pailler, F.Langlais, X.Bourrat, Boron-bearing species in ceramic matrix composites for long-term aerospace applications, *J. Sol. St. Chem.* **177**, 449-56 (2004).
[5]L. Vandenbulcke, G. Fantozzi, S. Goujard, M. Bourgeon, Outstanding ceramic matrix composites for high temperature applications, *Adv. Eng. Mater.* **7**, 137-142 (2005).
[6]A.O.Sezer, J.I.Brand, Chemical vapor deposition of boron carbide, *Mater. Sci. Eng. B* **79**, 191-202 (2001).
[7]V.Cholet, R.Herbin, L.Vandenbulcke, Chemical vapour deposition of boron carbide from BBr$_3$-CH$_4$-H$_2$ mixtures in a microwave plasma, *Thin Solid Films* **188**, 143-55 (1990).
[8]L. Vandenbulcke, G. Vuillard, CVD of Amorphous Boron on Massive Substrates, *J. Electrochem. Soc.* **124**, 1931-7 (1997).
[9]S. Goujard, L. Vandenbulcke, C. Bernard, Thermodynamic study of the chemical vapor deposition in the Si-B-C-H- Cl system, *CalPhaD* **18**, 369-85 (1994).
[10]M.Karaman, N.A.Sezgi, T.Doğu, H.Ö. Özbelge, Kinetic Investigation of Chemical Vapor Deposition of B4C on Tungsten Substrate *AIChE J.* **52**, 4161-6 (2006).
[11]M.Karaman, N.A.Sezgi, T.Doğu, H.Ö. Özbelge, Mechanism Studies on CVD of Boron Carbide from a Gas Mixture of BCl$_3$, CH$_4$, and H$_2$ in a Dual Impinging-jet Reactor, *AIChE J.* **55**, 701-9 (2009).
[12]J.Berjonneau, G.Chollon, F.Langlais. Deposition process of amorphous boron carbide from CH$_4$/BCl$_3$/H$_2$ precursor, *J. Electrochem. Soc.* **153**, C795-C800 (2006).
[13]J. Berjonneau, F. Langlais, G. Chollon, Understanding the CVD process of (Si)-B-C ceramics through FTIR spectroscopy gas phase analysis, *Surf. Coat. Technol.* **201**, 7273-85 (2007).
[14]J. Berjonneau, G. Chollon, F. Langlais, Deposition process of Si-B-C ceramics from CH$_3$SiCl$_3$/BCl$_3$/H$_2$ precursor *Thin Solid Films* **516**, 2848 (2008).
[15]G. Chollon, F. Langlais, J. Berjonneau, Gas Phase Deposition and Characterization of (Si)-B-C Ceramics, *Electrochem. Soc. Trans.* **25**, 15-21 (2009).
[16]H. B. Schlegel, S. J. Harris, Thermochemistry of BH$_m$Cl$_n$ calculated at the G-2 level of theory, *J. Phys. Chem.* **98**, 11178-80 (1994).
[17]M. D. Allendorf, C. F. Melius, Thermochemistry of Molecules in the B–N–Cl–H System: Ab Initio Predictions Using the BAC–MP4 Method, *J. Phys. Chem. A* **101**, 2670-80 (1997).
[18]Y. Zeng, K. Su, J. Deng, T. Wang, Q. Zeng, L. Cheng, L. Zhang, Thermodynamic investigation of the gas-phase reactions in the chemical vapor deposition of boron carbide with BCl$_3$–CH$_4$–H$_2$ precursors, *J. Mol. Struct. (THEOCHEM)* **861**, 103-16 (2008).
[19]T. Wang, K. Su, J. Deng, Y. Zeng, Q. Zeng, L. Cheng, L. Zhang, Reaction Thermodynamics In Chemical Vapor Deposition Of Boron Carbides With BCl$_3$–C$_3$H$_6$ (Propene)-H$_2$ Precursors, *J. Theor. Comput. Chem.* **7**, 1269-312 (2008).
[20]S. Harris, J. Kiefer, Q. Zhang, A. Schoene, K.-W. Lee, Reaction mechanism for the thermal decomposition of BCl$_3$/H$_2$ gas mixtures, *J. Electrochem. Soc.* **145**, 3203-11 (1998).
[21]I. Lengyel, K. F. Jensen, A chemical mechanism for in situ boron doping during silicon chemical vapor deposition, *Thin Solid Films* **365**, 231-41 (2000).
[22]S. Zhang, Y. Zhang, C. Wang, Q. Li, Density functional theory and ab initio direct dynamics study on the reaction of BCl$_3$ + H → BCl$_2$ + HCl, *Chem. Phys. Lett.* **373**, 1-7 (2003).
[23]Y. Qi, X.-F. Chen, K.-L. Han, Direct Dynamics Investigation on Mechanism of Reaction between Trichloride and H Radical, *J. Theor. Comput. Chem.* **5**, 51-7 (2005).
[24]G. Reinisch, Theoretical and experimental studies of the chemical vapor deposition of carbides, Ph.D. Thesis N° 4038, University Bordeaux, 2010. (in French)
[25]M. J. Frisch, G. W. Trucks, H. B. Schlegel, G. E. Scuseria, M. A. Robb, J. R. Cheeseman, J. A. Montgomery, Jr., T. Vreven, K. N. Kudin, J. C. Burant, J. M. Millam, S. S. Iyengar, J. Tomasi, V. Barone, B. Mennucci, M. Cossi, G. Scalmani, N. Rega, G. A. Petersson, H. Nakatsuji, M. Hada, M.

Ehara, K. Toyota, R. Fukuda, J. Hasegawa, M. Ishida, T. Nakajima, Y. Honda, O. Kitao, H. Nakai, M. Klene, X. Li, J. E. Knox, H. P. Hratchian, J. B. Cross, V. Bakken, C. Adamo, J. Jaramillo, R. Gomperts, R. E. Stratmann, O. Yazyev, A. J. Austin, R. Cammi, C. Pomelli, J. W. Ochterski, P. Y. Ayala, K. Morokuma, G. A. Voth, P. Salvador, J. J. Dannenberg, V. G. Zakrzewski, S. Dapprich, A. D. Daniels, M. C. Strain, O. Farkas, D. K. Malick, A. D. Rabuck, K. Raghavachari, J. B. Foresman, J. V. Ortiz, Q. Cui, A. G. Baboul, S. Clifford, J. Cioslowski, B. B. Stefanov, G. Liu, A. Liashenko, P. Piskorz, I. Komaromi, R. L. Martin, D. J. Fox, T. Keith, M. A. Al-Laham, C. Y. Peng, A. Nanayakkara, M. Challacombe, P. M. W. Gill, B. Johnson, W. Chen, M. W. Wong, C. Gonzalez, and J. A. Pople, Gaussian 03, revision C.02, Gaussian, Inc., Wallingford CT, 2004..

[26]K. F. Jensen, Introduction to Computational Chemistry, 2nd ed., Wiley, New York, 2007.

[27]L. A. Curtiss, K. Raghavachari, P. C. Redfern, V. Rassolov, J. A. Pople, Gaussian-3 (G3) theory for molecules containing first and second-row atoms, *J. Chem. Phys.* **109**, 7764-76 (1998).

[28]B. Anantharaman, C. F. Melius, Bond additivity corrections for G3B3 and G3MP2B3 quantum chemistry methods, *J. Phys. Chem. A* **109**, 1734-47 (2005).

[29]G. Reinisch, J.-M. Leyssale, G. L. Vignoles, Hindered rotor models with variable kinetic functions for accurate thermodynamic and kinetic predictions, *J. Chem. Phys.* **133**, 154112 (2010).

[30]K. S. Pitzer, W. D. Gwinn, Energy levels and thermodynamic functions for molecules with internal rotation. I. Rigid frame with attached tops, *J. Chem. Phys.* **10**, 428-40 (1942).

[31]J. E. Kilpatrick, K. S. Pitzer, Energy levels and thermodynamic functions for molecules with internal rotation. III Compound rotation. *J. Chem. Phys.* **17**, 1064-75 (1949).

[32]J. Pfaendtner, X. Yu, L. J. Broadbelt, The 1-D hindered rotor approximation, *Theor. Chem. Acc.* **118**, 881-98 (2007).

[33]D. R. Herschbach, H. S. Johnston, K. S. Pitzer, and R. E. Powel, Theoretical pre-exponential factors for 12 bimolecular reactions, *J. Chem. Phys.* **25**, 736-41 (1956).

[34]D. G. Truhlar, B. C. Garrett, and S. J. Klippenstein, Current Status of Transition-State Theory, *J. Phys. Chem.* **100**, 12771-800 (1996).

[35]S. J. Klippenstein, An Efficient Procedure for Evaluating the Number of Available States within a Variably Defined Reaction Coordinate Framework, *J. Phys. Chem.* **98**, 11459-64 (1998).

[36]Y. Georgievskii, S. Klippenstein, Variable reaction coordinate transition state theory: Analytic results and application to the $C_2H_3+H \rightarrow C_2H_4$ reaction, *J. Chem. Phys.* **118**, 5442-55 (2003).

[37]G. Reinisch, J.-M. Leyssale, G. L. Vignoles, A theoretical study of the decomposition of BCl_3 induced by a H radical, *J. Phys.Chem. A* **115**, 4786–4797 (2011).

[38]J. Jourdain, G. Laverdet, G. Lebras, J. Combourieu, Kinetic-Study By Electron-Paramagnetic-Resonance and Mass-Spectrometry of the Elementary Reactions of Boron-Trichloride with H and O Atoms and OH Radicals, *J. Chim. Phys.* **78**, 253-7 (1981).

[39]P. G. Bolhuis, F. S. Csajka, D. Chandler, Transition Path Sampling and the Calculation of Rate Constants, *J.Chem.Phys.* **108**, 1964-77 (1998).

[40]C. Dellago, P. G. Bolhuis , P. L. Geissler, Transition Path Sampling, *Adv. Chem. Phys.* **123**, 1-78 (2002).

[41]K. Norinaga, O. Deutschmann, Detailed kinetic modeling of gas-phase reactions in the chemical vapor deposition of carbon from light hydrocarbons, *Ind. Eng. Chem. Res.* **46**, 3547-57 (2007).

[42]G. Reinisch, J.-M. Leyssale, G. L. Vignoles, Reaction Mechanism for the Thermal Decomposition of $BCl_3/CH_4/H_2$ Gas Mixtures, *J. Phys. Chem. A* **115**, 11579-88 (2011).

[43]D. G. Goodwin, An Open-Source, Extensible Software Suite for CVD Process Simulation *ECS Procs* **2003-08**, 155-162 (2003).

[44]G. Reinisch, J.-M. Leyssale, N. Bertrand, G. Chollon, F. Langlais, G. L. Vignoles, Experimental and theoretical investigation of BCl_3 decomposition in H_2, *Surf. Coat. Technol.* **203**, 643-7 (2008).

[45]G. Reinisch, S. Patel, G. Chollon, J.-M. Leyssale, D. Alotta, N. Bertrand, G. L. Vignoles, Methyldichloroborane evidenced as an intermediate in the CVD synthesis of boron carbide, *J. Nanosci. Nanotech.* **11**, 8323-27 (2011).

IMAGE-BASED 2D NUMERICAL MODELING OF OXIDE FORMATION IN SELF-HEALING CMCS

V. Dréan[1], G. Perrot[1,2], G. Couégnat[2], M. Ricchiuto[1], G. L. Vignoles[2]

1. INRIA Sud-Ouest
350 Cours de la Libération,
F33410 TALENCE Cedex, France

2. University Bordeaux 1
LCTS – Lab. for ThermoStructural Composites
3, Allee La Boétie
F33600 PESSAC, France

ABSTRACT – Excellent lifetimes make Self-healing Ceramic-Matrix Composites (CMCs) promising candidates for jet engine hot parts. These composites have a 3D arrangement of SiC fiber tows infiltrated by a multilayer matrix. A pyrocarbon interphase acts as a crack deviator, SiC matrix layers bring stiffness, and boron-containing phases are able to produce above 450°C a glass oxide preventing further oxidation by a diffusion barrier effect. This paper introduces an image-based numerical simulation of the self-healing mechanism under oxygen gas. Existing 0D or 1D models give the time evolution of the oxygen concentration in the weakest fiber and deduce from it a global lifetime through an oxygen-controlled slow crack growth rate law. We propose an approach in which the resolution domain is a 2D FE mesh directly obtained from transverse images of a tow containing the crack. Oxygen diffusion, carbon consumption around the fibers, and conversion of boron carbide into boron oxide are simulated. The model solves mass balances in terms of heights of oxygen (gaseous or dissolved), liquid oxide, pyrocarbon, and boron-containing phase. All the heights are considered perpendicular to the image plane (thin layer approximation), and represent the unit volume (per square meter) occupied by each phase. Preliminary results on images containing several dozens of fibers and a multilayer matrix are discussed.

INTRODUCTION

Self-healing Ceramic-Matrix Composites (SH-CMCs) are promising candidate materials for civil aircraft jet engine hot parts[1]. Tests in actual conditions of use have been performed, showing excellent lifetimes[2]. These materials have a woven/interlocked 3D arrangement of SiC fiber tows, infiltrated by a multilayer matrix. The interphase and matrix layers all have a specific role. First, there is a pyrocarbon interphase acting as a crack deviator and preventing premature fiber failure while the matrix undergoes multiple cracking, that is, progressive damage. Second, SiC matrix layers bring stiffness and are chemically inert at moderate temperatures. Finally, boron-containing phases produce above 450°C a liquid oxide which fills cracks and prevents further oxidation by a diffusion barrier effect[3], thus increasing by large amounts the material lifetime during high-temperature oxidation[4].

The massive production of SH-CMC parts in engines can only be envisaged if there is a sufficient confidence in the material lifetime duration. The trouble is that the material can last as long as ten years; hence, a material development cycle based only on experimental characterization is not feasible. Modeling is mandatory in order to incorporate as much understanding of the material's physics as possible, and identify and explain the experiments that have been carried out in "accelerated degradation" conditions, or on material parts. The present work has been developed in this frame.

The material lifetime results from a balance between degradation and self-healing[5]. In the SH-CMC, the network of matrix cracks brings the oxygen in contact with all the inner material constituents; they are transformed gradually into oxides, some of them being gaseous (*e.g.* carbon),

117

liquid (*e.g.* boron) or solid (*e.g.* silica at moderate temperatures). The calculation of the chemical reaction rate everywhere in the material is a key point for lifetime prediction. Past reports have described models which give the time evolution of the oxygen concentration in, say, some key part of the composite (i.e. the weakest, most exposed fiber); they deduce from this quantity a global lifetime because the tensile strength of this fiber can be related to oxygen exposure through an oxygen-dependent slow crack growth law[6,7]. These models were developed in 0 [8,9] or 1 [10] space dimensions. In order to gain further insight in the self-healing phenomenon at fiber and tow scale, we propose now an image-based approach, in which the resolution domain is a 2D FE mesh directly obtained from original mesh generation software. Though the whole history of the material includes liquid phase flow, we will deliberately discard this phenomenon here and focus on the first steps of the material evolution, that is, the formation of the liquid oxide and the plugging of the crack.

The paper is organized as follows. First, we will recall the main physicochemical phenomena that are being accounted for in the model; then, model equations are built, and an analysis of their stiffness is presented. The next part describes the numerical implementation. Finally, some results are presented and discussed.

MATERIAL STRUCTURE AND PHYSICO-CHEMICAL PROCESSES

The modeled SH-CMCs are depicted on fig. 1a. Several domains are easily recognized:

- 1. The SiC fibers, which are sensitive to oxidation through a subcritical crack growth behavior. Their diameter is in the range 8-15 μm.

- 2. The thin pyrocarbon (PyC) interphase coating the fibers. This layer acts a crack deviator, helping the development of multi-cracking, which gives the material its exceptional mechanical behavior[11,12]. In this work we will consider that cracks are perpendicular to the fiber and are limited by the fiber boundary. In particular, fig. 1b illustrates the fact that in our model the fibers are bridging the cracks.

- 3. The multi-layer matrix containing two constituents: silicon carbide, which is nearly inert at the temperatures typical of civil aircraft applications (it begins to oxidize appreciably above 900°C in dry air [13,14]), and boron carbide, yielding a B_2O_3 glass phase upon oxidation[3,4].

- 4. A number of cusp-shaped pores left by the production process. In particular, the PyC, SiC and boron carbide layers are deposited by Chemical Vapor Infiltration (CVI)[15,16], a process that gives characteristic continuous layers growing in conformal shape with respect to the initial substrate (*i.e.* the fibers). The layer thicknesses are larger in the outer parts of the fiber bundles or tows, because of the competition between chemical deposition reactions and gas diffusion[17]. As a result, some pores remain between the fibers after infiltration. These imperfections may or may not be connected to the outer gas phase, depending on whether the crack network intercepts them.

Concerning the material evolution, the dominating phenomenon is oxidation. Accordingly, oxygen balance is the natural ingredient of the model, as well as balances for all oxidized materials, namely, pyrocarbon - undergoing active oxidation and giving gaseous CO_2 - and boron carbide - yielding a liquid B_2O_3 oxide - together with dissolved CO_2. Silicon carbide may also produce a solid silica layer upon oxidation; however, the temperatures considered here (< 650°C) are not high enough to allow the development of a significant silica thickness. Moreover, here we only address the case in which the boron oxide liquid is viscous enough not to flow significantly away from the position where it has been created.

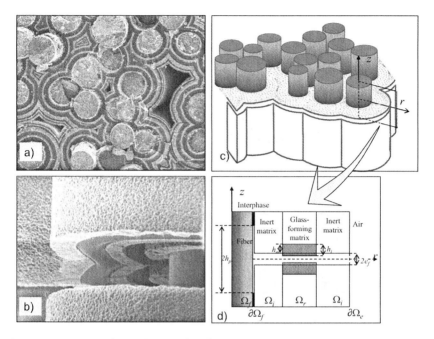

Figure 1. a) Structure of a multilayer self-healing ceramic-matrix composite viewed in a transverse direction with respect to fibers b) Crack in the multilayer matrix with a bridging fiber[18]; c) simulation domain viewed in elevation, d) detail of the simulation domain subsets and of several parameters.

MODEL SETUP

In order to describe the formation of the protective layer, oxygen diffusion, carbon consumption around the fibers, and conversion of boron carbide into boron oxide are simulated. The underpinning model consists of mass balances of oxygen (gaseous or dissolved), boron carbide, boron oxide, and pyrocarbon. The liquid oxide obtained as a product is assumed viscous enough not to flow away at least in the initial phase of the healing process simulated here. As a consequence, its growth is vertical above the carbide/oxide interface (see fig. 1d). Using a thin layer approximation, the mass balances are written in terms of heights representing the unit volume (per square meter) of certain local surface element. All the heights are considered perpendicular to the plane of the image, assumed to be the crack plane. In particular, the simulation domain consists of several subsets. As depicted by figs. 1c and 1d we can distinguish the following sub-domains:

- the inert matrix domain (silicon carbide) Ω_i;
- the reactive matrix domain (boron carbide) Ω_r;
- the fibers boundary $\partial\Omega_f$, modeled as an infinitely thin one-dimensional reactive layer of pyrocarbon ;
- the pore boundaries $\partial\Omega_e$, subjected to exposure to external oxygen ;
- the fibers Ω_f, bridging the crack, which are not part of the computational domain.

The chemical balances of the oxidation processes are:

$$B_4C_{1.6}(s) + 4.6\ O_2(g) \rightarrow 2\ B_2O_3(l) + 1.6\ CO_2(g) \tag{1}$$

$$C(s) + O_2(g) \rightarrow CO_2(g) \tag{2}$$

Note that the boron carbide in reaction (1) contains excess carbon with respect to the usual stoichiometry B_4C, as found in CVD experiments[19,10] – we will note it B_xC in the following. The mass balance equations are considered as averaged over the direction perpendicular to the crack. This implies that several heights involved in the problem become variables defined over some part of the resolution domain. They are:

- the height h_m of consumed boron carbide, defined only on Ω_r;
- the height h_l of liquid boron oxide, which is larger than h_m because of the molar volume differences (*i.e.* B_2O_3 has a larger molar volume than B_xC), defined on Ω_r;
- the height of consumed pyrocarbon h_p, existing only on the fiber/matrix interface $\partial\Omega_{fi}$ (considering that its thickness is so small that it does not need to be meshed).

The balance equations for oxygen, boron carbide and boron oxide involve source terms describing diffusion-limited oxidation reactions. Pyrocarbon oxidation appears as a boundary condition on the fiber surface for the oxygen and pyrocarbon balances. The oxygen mass balance reads:

$$\partial_t\left(h_g C_{O_2}\right) + \nabla\cdot\left(-h_g D_{O_2}\nabla C_{O_2}\right) = -\Phi_r\frac{k_l}{4.6}C_{O_2} \quad \text{in } \Omega_r \text{ and } \Omega_i \tag{3}$$

$$-h_g D_{O_2}\nabla C_{O_2}\cdot n_{fi} = -k_p e_p C_{O_2} \quad \text{on } \partial\Omega_{fi} \tag{4}$$

$$\begin{cases} -D_{O_2}\nabla C_{O_2}\cdot n_{ei} = -\dfrac{D_{O_2}^g}{\delta}\left(C_{O_2} - C_{O_2}^{ext}\right) \\[2mm] \qquad\qquad or \\[2mm] C_{O_2} - C_{O_2}^{ext} = 0 \end{cases} \quad \text{on } \partial\Omega_{ei} \tag{5}$$

In eqs. (3-4) we have used the variable h_g which is the available height for oxygen gas diffusion, and which is given by $h_g = e_f + \Phi_r(h_m - h_l)$, where Φ_r indicates whether one lies or not in the reactive phase region. Eq. (5) displays two possible kinds of boundary conditions: either of the von Neumann type (similar to Fourier boundary condition in heat balances), or of the Dirichlet type (fixed assigned concentration). The time variations of the various heights involved in the system are given by the following ODEs:

$$\partial_t\left(h_g\right) = \Phi_r\left(\upsilon_{B_xC} - 2\upsilon_{B_2O_3}\right)\frac{k_l}{4.6}C_{O_2} \quad \text{in } \Omega_r \tag{6}$$

$$\partial_t\left(h_l\right) = \Phi_r\frac{2\upsilon_{B_2O_3}k_l}{4.6}C_{O_2} \quad \text{in } \Omega_r \tag{7}$$

$$\partial_t\left(h_p\right) = \upsilon_{PyC}k_p C_{O_2} \quad \text{on } \partial\Omega_{fi} \tag{8}$$

The $\dot{h}$ term represents the possibility of a time variation in the crack opening, due for instance to variations in mechanical loading. The source terms for pyrocarbon and boron carbide oxidation depend on a reaction rate that follows a linear-parabolic law, as in the Deal-Grove model[20]:

$$k_l = \left(\frac{1}{k_{BxC}} + \frac{h_l}{D_{O_2}^l}\right)^{-1} \tag{9}$$

$$k_p = \left(\frac{1}{k_{PyC}} + \frac{h_p}{D_{O_2}^g}\right)^{-1} \tag{10}$$

Though not very necessary on a physical point of view (because k_{B_xC} is very large in front of $D_{O2}{}^l/h_l$), eq. (10) ensures a proper initialization of the numerical resolution. The initial conditions are:

$$C_{O_2}(t=0) = C_{O_2}^{ext}; h_{O_2} = e_f; h_l = 0; h_p = h_g = 1 \qquad (11)$$

The remaining notation is explained in Table 1, where typical values for all coefficients are also reported.

Table 1. Meaning and units of the symbols used in the model.

Symbol	Meaning	Value	Unit
C_{O2}	Oxygen concentration		mol.m^{-3}
D_{O2}	Oxygen diffusion coefficient		m^2.s^{-1}
$D_{O2}{}^g$	Oxygen diffusion coefficient	$9.4 \cdot 10^{-5}$	m^2.s^{-1}
$D_{O2}{}^l$	Oxygen diffusion coefficient	$5.4 \cdot 10^{-10}$	m^2.s^{-1}
e_f	Crack opening	10^{-6}	m
e_p	Pyrocarbon interphase thickness	10^{-7}	m
h_g	Available height for gaseous transfer		m
h_l	Liquid phase height		m
h_m	Consumption height of B_xC		m
h_p	Consumption height of pyrocarbon		m
k_{PyC}	Rate constant for PyC consumption	$1.7 \cdot 10^{-3}$	m.s^{-1}
k_p	Effective rate constant for PyC consumption		m.s^{-1}
k_{BxC}	Rate constant for B_xC consumption	$1 \cdot 10^{-3}$	m.s^{-1}
k_l	Effective rate constant for B_xC consumption		m.s^{-1}
υ_{B2O3}	Boron oxide molar volume	$4.492 \ 10^{-5}$	m^3.mol^{-1}
υ_{BxC}	Boron carbide molar volume	$2.314 \ 10^{-5}$	m^3.mol^{-1}
υ_{PyC}	Pyrocarbon molar volume	$0.7059 \ 10^{-5}$	m^3.mol^{-1}
δ	Boundary layer thickness	$2 \ 10^{-4}$	m
Φ_r	Reactive phase indicator (1 in Ω_r, 0 elsewhere)		-

NUMERICAL RESOLUTION

The resolution domain is obtained from image processing routines. Two options are possible: (i) a routine allows extracting from a micrograph of a transverse slice of a given tow a finite element mesh, with a proper description of the domains as listed above; (ii) another routine allows extracting from a micrograph information on the fiber size and neighborhood distribution, then synthesizing a random artificial medium with the same fiber properties, and finally creating the multilayer matrix by mesh dilation operations. In this way, it is possible to work either on an actual material, or on a virtual material.

The oxygen diffusion problem (3-5) has been discretized using a standard Galerkin approach on triangular P1 Lagrange finite elements. Time integration of the ODEs (6-8) is performed using an implicit backward differencing and solving the algebraic problem by means of standard Newton-Raphson iterations.

Using dimensional analysis, an estimate of the relative importance of the different phenomena has been carried out, based on the data reported in Table 1. As a result of this analysis, one can see that the time scale associated to oxygen diffusion in gas phase is small compared to the time scale of liquid oxide formation. Thus, the oxygen diffusion equation (3-5) may be considered as a decoupled steady problem to be used as a starting point to integrate the height evolution equations (6-8). This decoupled approach allows to first solve equations (3-5) for the oxygen concentration field with frozen height values; then, larger time steps are taken for the computation of an increase of the heights, assuming a steady concentration field. After one large time step, the new values for the heights are used to obtain a solution of the oxygen diffusion equation. The procedure is carried out until h_g reaches zero on some points of the mesh.

RESULTS

Illustrative results are presented on an SH-CMC tow image containing approximately 50 fibers. The fibers were obtained by scanning and thresholding a transverse micrograph of a CMC; the matrix layers were generated by mesh dilations. The layer sequence is (from fiber to surface) $PyC/SiC/B_xC/SiC/B_xC/SiC$. Symmetry boundary conditions were applied everywhere the material was meeting the image border. All pores were considered as equally connected to the exterior oxygen source; a Neumann boundary condition was applied.

Figure 2 gives shows the numerical results when plug formation is starting, at $t = 112$ s.

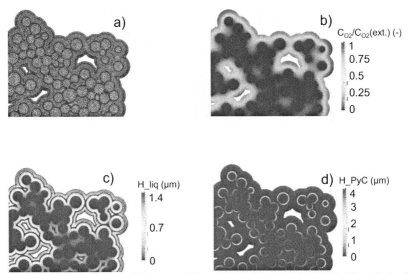

Figure 2. Visualization of the numerical results in a base case (see Table 1), at the beginning of healing ($t = 112$ s). a) Computational mesh, b) Scaled oxygen partial pressure field, c) glass oxide height, d) Consumed pyrocarbon height.

Figure 2b clearly evidences the depletion of oxygen throughout the crack. Indeed, consumption by the glass-forming carbide and by carbon is strongly limited by supply of gas by diffusion. So, only the boron carbide layers located close enough to the boundaries have reacted, as seen on fig. 2c. In correlation with this, the consumed pyrocarbon heights are larger for the mesh points lying closer to the borders; a depletion effect is also seen for the points which lie close to a second fiber, as seen in fig. 2d. In fig. 3a are reported the consumed PyC heights as a function of the distance to the closest border, at $t = 32$s. Evidently, the mesh points lying closest to the surface are more oxidized; however one notes a large dispersion of the consumed height values everywhere in the material, as commented before on fig. 2d. To quantify this dispersion, the mesh points lying closest to the border (highlighted in fig. 3a) are studied and their histogram is plotted at figure 3b. The standard deviation is 14% of the average value, a rather appreciable amount.

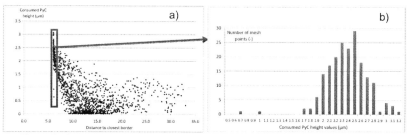

Figure 3. Consumed PyC height results at t = 32 s. a) Values at mesh points *vs.* distance to the closest border. b) Values histogram for the mesh points lying closest to the border.

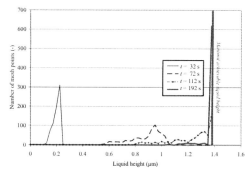

Figure 4. Glass height histograms at selected time values, for the mesh points lying closest to the border.

Two-dimensional effects, such as competition in oxygen consumption between adjacent fibers, can be dramatic, as shown by the histograms of glass height formation on Figure 3. As we can see, the spreading of the distribution is rather large at the beginning, then further spreads, before arriving at the maximal achievable height. When the surface barrier formation begins, *i.e.* when a first mesh point reaches the maximal achievable value for the liquid height (here, 1.35 μm), other surface points are still far from this state (minimum around 0.7 μm). Consequently, it takes a rather long time between the beginning and the end of barrier formation. The estimation in the present case is ~ 80 s, *i.e.* close to the beginning time itself (112 s); however it is certainly an over-estimation, and it can be corrected when taking into consideration the glass phase flow.

CONCLUSION AND OUTLOOK

This work focuses on the modeling of the first moments of the formation of a glassy boron oxide protective barrier in an SH-CMC tow. The originality is to carry out modeling in 2D in a crack plane supposed to be transverse with respect to fibers. The results show that it is possible to quantify the gradients of barrier formation throughout the section, and obtain an average and standard deviation of the self-healing behavior.

Future work is aimed to account for the flow of the glass phase and of the diffusion of oxygen through it, in order to describe the "latency" period of the self-healing composite.

ACKNOWLEDGEMENTS

This work has been funded by GIS "Advanced Materials in Aquitaine" through a post-doc grant to V. D.. The authors also wish to thank Pr. F. Rebillat (LCTS, University Bordeaux) for fruitful discussion on the model.

REFERENCES

[1]F. Christin, A Global Approach to Fiber Architectures and Self-Sealing Matrices: From Research to Production. *Int. J. Appl. Ceram. Technol.* **2**, 97-104 (2005).
[2]J.-C. Cavalier, I. Berdoyes, and E. Bouillon, Composites in aerospace industry, *Adv. Sci. Technol.* **50**, 153-62 (2006).
[3]S. Goujard, J.L. Charvet, J.L. Leluan, F. Abbé, and G. Lamazouade, French Patent N° 95 03606 (1995).
[4]P. Forio and J. Lamon, Fatigue behavior at high temperatures in air of a 2D SiC/Si-B-C composite with a self-healing multilayered matrix, *Ceram. Trans.* **128**, 127-41 (2001).
[5]L. Quémard, F. Rebillat, A. Guette, H. Tawil and C. Louchet-Pouillerie, Self-healing mechanisms of a SiC fiber reinforced multi-layered ceramic matrix composite in high pressure steam environments, *J. Eur. Ceram. Soc.*, **27-4**, 2085-94 (2007).
[6]W. Gauthier, J. Lamon and R. Pailler, Static fatigue of monofilaments and of SiC Hi-Nicalon yarns at 500°C and 800°C, *Rev. Compos. Matér. Avancés* **16**(2), 221-41 (2006). (in French)
[7]P. Ladevèze and M. Genet, A new approach to the subcritical cracking of ceramic fibers, *Compos. Sci. Technol.*, **70**, 1575-83 (2010).
[8]C. Cluzel, E. Baranger, P. Ladevèze, and A. Mouret, Mechanical behaviour and lifetime modelling of self-healing ceramic matrix composites subjected to thermomechanical loading in air, *Compos. Part A* **40**(8), 976-84 (2009).
[9]L. Marcin, E. Baranger, P. Ladevèze, M. Genet and L. Baroumes, L., Procs 7th Intl. Conf. on High Temperature Ceramic Matrix Composites (HT-CMC7), Bayreuth, Germany, p. 194-202 (2010).
[10]F. Rebillat, Original 1D oxidation modeling of composites with complex architectures, Procs 5th Intl. Conf on High Temperature Ceramic Matrix Composites (HT-CMC5). Seattle, WA, p. 315-20 (2005).
[11]V. Dréan, M. Ricchiuto and G. L. Vignoles, Two-dimensional oxydation modelling of MAC composite materials, INRIA Reports N°7417 & 7418 (2010)(in French).
[12]G. Camus, Modeling of the Mechanical behavior and damage process of fibrous ceramic matrix composites: application to a 2-D SiC/SiC, *Intl. J. Sol. Struct.* **37**(6), 919-42 (2000).
[13]F. Lamouroux, G. Camus and J. Thébault, Kinetics and Mechanism of Oxidation of 2D Woven C/SiC Composites: I, Experimental Approach, *J. Amer. Ceram. Soc.* **77**(8), 2049-57 (1994).
[14]F. Lamouroux and G. Camus, Oxidation Effects on the Mechanical Properties of. 2D Woven C/SiC Composites, *J. Eur. Ceram. Soc.***14**(2), 177-88 (1994).
[15]R. Naslain and F. Langlais, Fundamental and practical aspects of the chemical vapor infiltration of porous substrates, *High Temp. Sci.* **27**, 221-35 (1990).
[16]T. M. Besmann, B. W. Sheldon, R. A. Lowden and D. P. Stinton, Vapor-phase fabrication and properties of continuous-filament ceramic composites, *Science* **253**, 1104-9 (1991).
[17]G. L. Vignoles, Modelling of CVI Processes, *Adv. Sci. Technol.* **50**, 97-106 (2006).
[18]X. Martin, Oxidation/Corrosion of $(SiC_f/SiBC_m)$ self-healing ceramic matrix composites, PhD thesis, Bordeaux I University, n°2749, 2003 (in French)
[19]J.Berjonneau, G.Chollon, F.Langlais, Deposition process of amorphous boron carbide from CH4/BCl3/H2 precursor, *J. Electrochem. Soc.* **153**(12), C795-800 (2006)

[20]B. E. Deal, A. S. Grove, General relationship for the thermal oxidation of silicon, *J. Appl. Phys.* **36**(12), 3770-8 (1965)

COMBINING X-RAY DIFFRACTION CONTRAST TOMOGRAPHY AND MESOSCALE GRAIN
GROWTH SIMULATIONS IN STRONTIUM TITANATE: AN INTEGRATED APPROACH FOR
THE INVESTIGATION OF MICROSTRUCTURE EVOLUTION

Melanie Syha[a], Michael Bäurer[a], Wolfgang Rheinheimer[a], Wolfgang Ludwig[b], Erik M. Lauridsen[c],
Daniel Weygand[a] and Peter Gumbsch[a,d]

[a]Institute for Applied Materials, Karlsruhe Institute of Technology, Karlsruhe, Germany
[b]European Synchrotron Radiation Facility, Grenoble, France
[c]Risø DTU National Laboratory, Roskilde, Denmark
[d]Fraunhofer Institute for Mechanics of Materials IWM, Freiburg, Germany

ABSTRACT
        Motivated by the recently reported a growth anomaly in strontium titatate bulk samples[1] , the
microstructure of bulk strontium titanate has been investigated by an integrated approach comprising
conventional metallography, three dimensional X-ray diffraction contrast tomography (DCT)[2], and the
observation of pore shapes in combination with mesoscale grain growth simulations. The
microstructural evolution in strontium titanate has been characterized alternating ex-situ annealing and
high energy X-ray DCT measurements, resulting in three dimensional microstructure reconstructions
which are complemented by crystallographic orientations obtained from diffraction information. These
investigations allow to establish a correlation between grain morphology, orientation dependent grain
boundary properties and growth behavior in these highly anisotropic materials. Together with energy
and mobility data gathered in conventional metallographical analysis, they serve as input for a 3D
vertex dynamics model[3].

INTRODUCTION
        Microstructural evolution as observed during heat treatments or densification processes has a
huge impact on thermal, electrical and mechanical properties of polycrystalline materials in general
and highly anisotropic ceramic materials, such as strontium titanate[4], in particular. Therefore, a
thorough understanding of the processes occurring during grain coarsening and densification is crucial.
Due to annealing temperatures of more than 1400°C needed to activate grain boundary motion in
strontium titanate obervable by X-ray tomography imaging, in-situ 3D characterizations of these
processes are not possible. Conventional metallography by sectioning the material however is
destructive and thus inhibiting the possibility to analyze the same microstructure multiple times at
different coarsening stages. The application of advanced X-ray imaging, in the form of diffraction
contrast tomography[2] allows to alternate ex-situ heat treatments at high temperatures and tomographic
scans, resulting in a step by step characterization of the structural evolution. The reconstructed
microstructures are then analyzed regarding interface orientation distribution functions of grain
ensembles, which is a signature of the anisotropy of the interfacial energy, related to the known
anisotropies of the surface energy of strontium titanate [5].
These three dimensional experimental characterizations provide realistic starting structures for the
three dimensional mesoscale model[3] and allow for a direct comparison of morphological, topological
and crystallographic quantities to data resulting from the simulation of grain growth in anisotropic
materials. Both conventional metallography and tomography data have been analyzed to derive more
realistic input data for the simulation of strontium titanate .

METHODS

The used methods both experimental setup and computational model shall be presented briefly. The microstructure evolution in a polycrystalline strontium titanate specimen has been characterized experimentally using SEM and X-Ray Diffraction Contrast Tomography (DCT). To access the evolution of the microstructure, the sample was annealed for 1h at 1600°C in between repetitive tomographic scans (details on the technique see below). For the simulations, a three dimensional vertex dynamics model (details below) was adapted for simulating strontium titanate. In a first step grain coarsening was simulated using experimentally observed energy landscapes.

X-ray Diffraction Contrast Tomography

The applied X-ray Diffraction Contrast Tomography technique uses synchrotron radiation for the non-destructive characterization of polycrystalline materials. During measurements, cylindrical samples are rotated and exposed to a monochromatic high energy X-ray beam while absorption as well as diffraction information is collected in 0.05° increments on the detector positioned behind the sample (figure 1). Full 360° scans of the sample are performed in order to be able to make use of Friedel pairs during the reconstruction procedure[6]

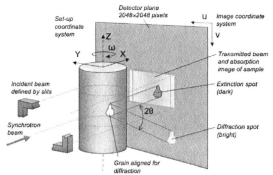

Figure 1. Experimental Set-Up during acquisition of absorption and diffraction information[2].

The reconstruction of both position and grain shape is primarily based on the exploitation of the diffraction spots gathered during these scans. After applying a standard algebraic reconstruction procedure to the groups of diffraction spots that have been identified to belong to the same grain, all grains are subjected to a uniform dilation procedure in order to render the structure volume filling. Since the sintered bulk material contains a small amount of porosity, complementary information on their distribution throughout the sample was gathered separately using phase contrast tomography scans. The collective pore ensemble was then subtracted from the volume filling three dimensional microstructure reconstruction. Crystallographic orientations for all grains are directly accessible via the Bragg angles, so that local interface orientation information and misorientation between grains can easily be derived.

Sample material

All specimens presented in this paper were manufactured from sintered strontium titanate raw material. The powders were prepared by the mixed oxide route from $SrCO_3$ and TiO_2 (both 99.9+%, Sigma Aldrich Chemie, Taufkirchen, Germany) using a molar Sr/Ti ratio of 0.996. After several milling and calcining steps, green bodies were prepared pressing the powders uniaxially to obtain the cylindrical shape and isostatically to increase the green density. Sintering these specimen at 1600 °C in oxygen atmosphere yield the final material with an average grain diameter of about 28μm (measured by means of equivalent circle diameter on optical micrographs) and a density of 98%. Fabrication process and metallographic characterizations of the microstructure were published in [2] and [7].

For the tomography samples, the final cylindrical shape with a diameter of roughly 300μm was obtained grinding conventional TEM samples using abrasive paper and a turning lathe.

Mesoscale Grain Growth Model

The vertex dynamics model is used to simulate the microstructure evolution in strontium titanate. In this particular approach, the microstructure is represented based on discretized interfaces only (figure 2) and the microstructural evolution is governed by minimization of the total interface energy, as detailed in [6].

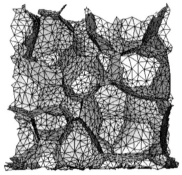

Figure 2. View into three dimensional structure of discretized interfaces, taken from 3D grain growth simulations using the vertex dynamics model described in [6].

For the present investigations, the model handles misorientation and inclination dependent grain boundary properties by assigning discrete interface energy and mobility values to triangles representing the smallest discretization unit of the grain boundary network. The orientation dependent surface energy functional of strontium titanate at 1400° C as measured by Sano et al. [5] (figure 3) was used to mimic an orientation dependent interface energy as an average of the contributions from the two adjacent grains with respect to their crystallographic orientations, as suggested by Saylor [8].

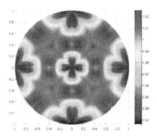

Figure 3. Surface energy as function of surface normal orientation for strontium titanate at 1400°C [5].

RESULTS

X-ray Diffraction Contrast Tomography

Figure 4 shows a three dimensional reconstruction and a cross-section of the aforementioned tomography sample before the annealing step.

Figure 4. Three dimensional reconstruction and cross-section through the strontium titanate microstructure before annealing. The three dimensional structure is colored randomly, while the cross-section is colored according to crystallographic orientation.

Grain shapes and pore location look conceivable and the average grain diameter of 28.6µm is in good agreement with results from equivalent circle diameter measurements on conventional micrographs of identical material (28,2µm), as is the grain size distribution[9]. The misorientation distribution resembles a perfect McEnzie plot as expected for a polycrystal with randomly oriented grains.

In figure 4, the apparent grain shapes in the cross-section of the sample are often delimited by rather straight grain boundaries. The true three dimensional shape reveals a trend for flat grain boundaries, exhibiting a preferred local interface inclination within the regarded grains local coordinate system (figure 5). The preferred orientation is the {100} direction.

Figure5. Two single grains extracted out of the pre-annealing structure triangulated and colored according to local interface orientation. The black areas indicate triangles along triple lines, e.g. each face of a grain is delimited by black triangles.

A comparison of reconstructions before and after annealing clearly monitors microstructure evolution featuring both growing and shrinking grains (figure 6). The two cross-sections are taken at approximately the same position within the sample. Coloring according to crystallographic orientation helps re-identifying corresponding grains. The black areas in figure 6 represent pores, which are found to be mainly located along grain boundaries and triple lines. The volume fraction of pores contained in

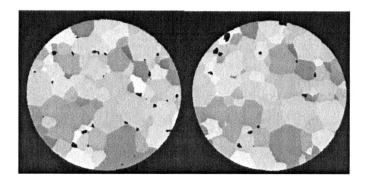

Figure 6. Cross sections at the same height of the reconstructions before (left) and after (right) annealing (Coloring according to crystallographic orientation of the grains).

the material decreases slightly, giving evidence of densification processes in the material. Interface orientation distributions of selected grains before and after annealing reveal an increasing number of {100} oriented interfaces. Qualitatively, these {100} oriented faces tend to flatten during microstructure evolution (figure 7).

Pore shapes in strontium titanate samples sintered at different temperatures were investigated in order to determine temperature and orientation dependent relative surface energies for the observed facets. The SEM image in figure 8(a) shows the shape of a pore. The orientation of the visible facets were measured by EBSD and relative area fraction of distinct facets have been extracted, allowing to estimate relative surface energies for the different occurring facets by comparison to a calculated Wulff shape (figure 8(b)).

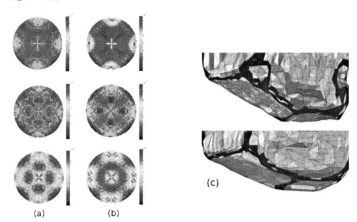

(a)   (b)

Figure 7. Stereographic plot of interface orientation distribution of three grains before (a) and after (b) annealing, {100} orientations are to be found in the center and on the outermost ring of the plot at 0°, 90°, 180° and 270°, respectively. Closeup of {100} oriented region of a selected grain before (c, upper image) and after (c, lower image) heating.

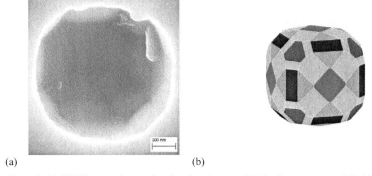

(a)   (b)

Figure 8: (a) SEM image of a pore in after sintering at 1600°C; (b) reconstructed Wulff shape.

Some initial investigations of the orientation dependency of interface mobilities have been conducted by measuring the migration distance of more than 3000 facets in the reconstructed microstructures that underwent evolution. These results have been compared to data collected during the growth of a single crystal with defined interface orientation into a polycrystalline matrix (annealing temperature 1600°C). The results are shown in table 1.

Table 1: Migration length (in μm) of different facets during annealing of the tomography samples (1h, 1600°C), tomography data and single crystal data normalized by their respective average migration length.

Facet	Migration length tomography [μm/h]	Normalized values obtained by tomography	Normalized values from single crystal SEM observation
100	3.20 +/- 1.5	0.98	0.78
110	3.53 +/- 1.5	0.99	1.08
111	4.20 +/- 1.5	1.18	1.10
310	3.30 +/- 1.5	0.93	1.03

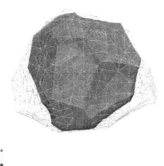

Figure 9. Selected grain at two different stages during the simulation (green: initial stage, wireframe: evolved stage). The coordinate system indicates the grains crystallographic orientation.

Vertex Dynamics Simulations

Grain growth simulations were performed starting from isotropically grown self similar grain structures using the orientation dependent interface energy landscape derived from the surface energy functional shown in figure 3. The influence of the interface orientation dependent functional can be observed both qualitatively and quantitatively: Figure 9 shows the evolution of a growing grain in the simulated structure, the green part representing the first time step and the wireframe the second time step. The {100} directions are indicated by the coordinate system printed in the lower left side of each picture. The grain shape becomes more cubic during the simulation and faces tend to rotate near to {100} orientations.

This is especially remarkable considering that a {100} orientation for the considered grain might be a high energy orientation with respect to the neighboring grains local coordinate system. Quantitatively, we see also a broadening of the angle distribution at the triple lines[6], resulting from the now non-zero gradient term in the Herrings relation[10].

The grain size distribution obtained from the reconstructed microstructures and optical micrographs are shown in figure 10. Both grain size distributions are best fitted by a log normal distribution (figure 10). Also, topological quantities (number of edges per grain, number of faces per grain, number of edges per face of a grain) from modeled grain ensembles using the orientation dependent interface energy functional and reconstructed tomography samples respectively are rather close(table 1).

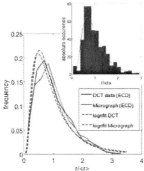

Figure 10: Grain size distributions as obtained by conventional metallography and tomography.

Table 2. Comparison of topological quantities as obtained by x-ray diffraction contrast tomography (DCT) and anisotropic vertex dynamics simulations.

	$\langle E_G \rangle$	$\langle F_G \rangle$	$\langle E_{GF} \rangle$
DCT	36.57	12.75	4.63
Vertex	35.35	13.84	4.98

The grain growth kinetics for both an isotropic and inclination dependent interface energy are shown in Figure 11a. The distribution of interface energies occurring within the structure is shown in figure 10b. During the evolution, a narrow peak develops, related to the energy cusp for the 100 orientation as visible in figure 3.

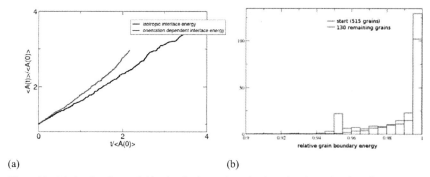

(a)   (b)

Figure 11: (a) simulated growth kinetics for isotropic and orientation dependent interface energies. (the average area is calculated as $\langle A(t) \rangle = \pi \langle r(t) \rangle^2$);(b) Distribution of interface energies in the structure at two stages during anisotropic coarsening starting from isotropically grown structure.

Figure 12 shows a grain structure imported to the simulation model. The voxel data has been translated to interface information The simulated structure has been taken from the pre-annealing stage of the tomography sample presented above it contains 37 grains and comprises about 10*6 triangles in the mesoscale grain growth model

Figure 12. 37 grains of the 3D reconstruction data imported in the vertex dynamics model.

DISCUSSION AND CONCLUSION
        Grain coarsening of strontium titanate was investigated by a combination of conventional metallography, mesoscale grain growth modeling and, more recently, 3D x-ray diffraction contrast tomography. The good agreement found in the topological quantities of reconstructed tomography structures and simulated grain boundary networks presented in table 2 confirms the modeling approach using the orientation dependent interface energy functional derived from the surface energies. Although the orientation dependent interface energy is averaged in order to account for both adjacent grains influence low energy directions seem to be preserved throughout the simulations (figure 11 (b)).

This goes together with an acceleration of the overall growth kinetics when switching from isotropic to anisotropic simulations using the inclination dependent interface energy functional (figure 11 (a)).

Results from both approaches indicate, that the growth kinetics of strontium titanate are (at least in the temperature regime considered in this study) strongly influenced by {100} interfaces. The grain boundary inclination distribution function also shows an increasing density of {100} oriented interfaces, indicating that the lower energy of {100} surfaces persists after joining and formation of the grain boundary.

Experiments designed to generate realistic input for the anisotropic grain growth model yield first orientation and temperature dependent surface energy/mobility values allowing for an improved interpolation of energy/mobility functionals for different temperatures to be used for upcoming studies. Despite large experimental uncertainties the migration length for distinct facets obtained by metallography and tomography measurements are of comparable size and the different facets appear in the same order.
Furthermore, first voxel based reconstructions can now be imported to the vertex dynamics model. These will serve in upcoming studies as starting structures for grain coarsening simulations under anisotropic conditions. Evolved structured may then be compared to the reconstruction of post-annealing stages of the samples and the results can be interpreted in the context of the experimentally investigated interface properties at different temperatures.

ACKNOWLEDGEMENTS
       This work was supported by the Deutsche Forschungsgemeinschaft (DFG) under Contract WE3544/4-1 and BA4143/2-1. We kindly acknowledge the Karlsruhe House of Young Scientists (KHYS) for covering travel expenses of one of the authors (M.S.), the European Synchrotron Radiation Facility (ESRF) for provision of beam time (MA-970) and Loredana Erra, beam line ID11, for technical support, and Peter Pfundstein, LEM, Karlsruhe Institute of Technology for EBSD measurements on pore shapes.

REFERENCES
[1]M. Bäurer, D. Weygand, P. Gumbsch and M. Hoffmann, Grain growth anomaly in strontium titanate , *Scripta Mat.*, **61** 584 (2009)
[2]G. Johnson, A. King, M.G. Honnicke, J. Marrow and W. Ludwig, X-ray diffraction contrast tomography: a novel technique for three-dimensional grain mapping of polycrystals. II. The combined case, *J. Appl. Cryst.* **41**, 310 (2008)
[3]M. Syha and D. Weygand, A generalized vertex dynamics model for grain growth in three dimensions, *Mod. Sim. Mat. Sci. Eng.*, **18**, 015010 (2010)
[4]S. Lee, W. Sigle, W. Kurtz and M. Rühle, Temperature Dependence of Faceting in SrTiO₃ Sigma 5 Bicrystal Grain Boundary, *Acta Mat.*, **51** 975 (2003)
[5]T. Sano, D.M. Saylor, and G.S. Rohrer, Surface Energy Anisotropy of SrTiO₃ at 1400 °C in Air, *J. Amer. Ceram. Soc.*, **86** , 1933 (2003)
[6]W. Ludwig, P. Reischig, A. King, M. Herbig, E.M. Lauridsen, G. Johnson, T. Marrow and J. Buffiere, Three-dimensional grain mapping by x-ray diffraction contrast tomography and the use of Friedel pairs in diffraction data analysis, *Rev. Sci. Instr.*. **80**, 33905 (2009)

[7]M. Bäurer, H. Kungl and M. Hoffmann, Influence of Sr/Ti Stoichiometry on the Densification Behavior of Strontium Titanate, *J. Amer. Ceram. Soc.,* **92** 601 (2009)

[8]D.M. Saylor, B.S. El-Dasher, Y. Pang, H.M. Miller, P. Wynblatt, A.D. Rollett, and G.S. Rohrer, "Habits of Grains in Dense Polycrystalline Solids," *J. Amer. Ceram. Soc.,* **87** 724-726 (2004).

[9]M. Syha, W. Rheinheimer, M. Bäurer, E.M. Lauridsen, W. Ludwig, D. Weygand, and P. Gumbsch, Three-dimensional grain structure of sintered bulk strontium titanate from X-ray diffraction contrast tomography, *Scripta Mat.* **66**, 1 (2011)

[10]C. Herring, Some Theorems on the Free Energies of Crystal Surfaces, *Phys. Rev.* **82**, 87-93 (1951)

CALCULATION OF GROWTH STRESS IN SiO$_2$ SCALES FORMED BY OXIDATION OF SiC FIBERS

R. S. Hay
Air Force Research Laboratory
Materials and Manufacturing Directorate, WPAFB, OH

ABSTRACT

A numerical method to calculate growth stress in SiO$_2$ scales formed during SiC fiber oxidation is described. Calculations were done for SiC fibers between 700° and 1300°C using previously measured Deal-Grove parameters for oxidation kinetics and an Eyring viscoplastic model for SiO$_2$ scale viscosity. Initial compressive stresses in SiO$_2$ of ~25 GPa from the 2.2× oxidation volume expansion are rapidly relaxed to lower levels by flow of silica with a shear stress-dependent viscosity. At 700° - 900°C, axial and hoop stress at the GPa level persist. Radial expansion of the outer scale causes hoop stress to become tensile; axial stress becomes tensile by the Poisson effect. Tensile hoop stresses can be >2 GPa for thick scales formed at <1000°C. Effects of different fiber radii on growth stresses are examined. Limitations of the method and analytical approximations are discussed.

INTRODUCTION

Constraint of the 2.2× volume expansion during oxidation of SiC to SiO$_2$ generates very large growth stresses. Microstructural evidence for these stresses exists for crystalline scales on SiC fibers. High dislocation densities in crystalline SiO$_2$ near the SiC-SiO$_2$ interface suggest high shear stresses exist during growth of new crystalline scale.[1] Axial cracks form in the outer scale from tensile hoop growth stress.[1-3] Tensile growth stress in the scale may decrease fiber strength by surface nucleated fracture.[4] Residual stresses affect SiC fiber strength and are important for SiC-SiC ceramic matrix composites (CMCs) for high temperature structural applications.[5-6]

Growth stress during silicon oxidation, which has a volume expansion similar to that for SiC oxidation, has been extensively modeled.[7-9] Most recent models recognize that flow at high stress is non-Newtonian (viscoelastic) and use the Eyring model for shear-stress dependent viscosity.[10-14] Radial compressive and tensile hoop growth stresses are predicted for oxidation of silicon fibers.[12,15-16] Axial stresses are generally ignored.[10-11,15,17] For structural fibers this is an important omission, because axial stress has the most significant effect on fiber strength. Applied tensile stress increases oxidation rates of silicon,[18-20] and recently this has also been demonstrated for SiC fibers.[21]

A method to calculate the radial, axial, and hoop growth stress components anywhere in a SiO$_2$ scale formed by oxidation of SiC fibers is presented. The method involves discretization of the scale into layers and separate calculation of stresses in each layer. Calculations are done at temperatures from 700° to 1300°C for amorphous scales. Growth stresses for SiC fibers with different radii are examined. Assumptions and limitations of the method are discussed. Complementary results, along with a more thorough description of the background information and method are given elsewhere.[22]

METHOD

General

A schematic of the volume expansion and attendant stresses during SiC fiber oxidation with discretization of SiO$_2$ scale into annular layers is in figure 1. There are two sources of stress. The first is the initial elastic constraint of the 2.2× volume expansion accompanying oxidation of SiC to SiO$_2$. This expansion is shown in stress-free (dilational) and constrained states (Fig. 1). The second is the circumferential expansion of old scale as it is radially displaced outward by formation of new scale (Fig. 1). This creates tensile hoop stress (σ_θ) in the outer scale,[4,23] and tensile axial stress (σ_z) by the Poisson effect.

The large volume expansion raises concerns for growth stress modeling. Elastic constraint of the 2.2× volume expansion causes ~25 GPa compressive stress in SiO$_2$ for SiC oxidation, which is much larger than stresses for which linear elasticity is valid.[7] However, high shear stresses relax very rapidly to values appropriate for linear elasticity if SiO$_2$ viscosity is shear-stress dependent, as in the Eyring model,[10-11,24-25] and discretization to small increments can confine incremental displacements and stress differences to values appropriate for linear theory.[13,26-27]

General models of oxidation growth stress are coupled diffusion-reaction and fluid-mechanical problems.[26] The oxygen diffusion rate drives the rate that stress in the oxide is generated. Stress, in turn, affects the oxygen diffusion rates. The measured oxidation rates for SiC, whether they are flat plates, fibers, or particles, inherently include the effect of growth stress; stress-free oxidation rates cannot be measured. The effect of stress on oxidation rate will therefore not initially be considered when modeling growth stress.

The following assumptions are used in modeling the growth stress:

1. Oxidation volume expansion is dilational.
2. Stresses resulting from constraint of oxidation expansion are relaxed by flow of SiO$_2$ with shear stress-dependent viscosity.
3. Discretization of oxidation to small increments allows use of linear elasticity.
4. Growth stress effects on oxidation kinetics are not considered.
5. Stress relaxation in the SiC fiber is negligible.

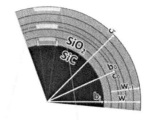
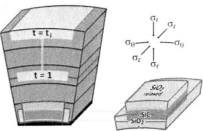

Fig. 1. *Schematic diagrams of oxidation of SiC to SiO$_2$. The original SiC surface is a blue line (c_o, b_o), midway through the scale thickness.*

Fiber Oxidation Kinetics

Fiber oxidation kinetics will not deviate significantly from flat-plate geometry kinetics until the oxidation product for 12 μm diameter fibers is several microns thick.[1,15,28] SiO$_2$ thickness (**w**) (Fig. 1) therefore obeys Deal-Grove kinetics for flat-plate geometry:[1,29]

$$w = \frac{1}{2}A\left[\sqrt{1 + \frac{4Bt}{A^2}} - 1\right] \qquad [1]$$

The parabolic and linear rate constants are **B** and **B/A**, respectively. **B** and **A** obey the usual Arrhenius relationships:

$$A = A_0 e^{\frac{-Q_A}{RT}} \qquad [2]$$

$$B = B_0 e^{\frac{-Q_B}{RT}} \qquad [3]$$

where **T** is absolute temperature, **R** is the gas constant, **Q$_A$** and **Q$_B$** are activation energies and **A$_o$** and **B$_o$** are pre-exponential factors. The SiC radius (**b**) after oxidation is (Fig. 1):

$$b = \sqrt{w^2\left(\Omega^2 - \Omega\right) + b_o^2} - \Omega w \qquad [4]$$

where **b$_o$** is the original fiber radius and **Ω** is the ratio of SiC/SiO$_2$ molar volume ratio. The outer radius of the SiO$_2$ scale (**c**) is (Fig. 1):

$$c = b + w \qquad [5]$$

Elastic Growth Stress
    The elastic stresses and strains are determined by modification of a method used for sequentially deposited coatings.[30] For discretized fiber oxidation, the "deposition" sequence is reversed; the last oxide increment forms at the SiC-SiO$_2$ interface, and the oldest is at the surface. The system is divided into the unoxidized SiC fiber, the SiO$_2$ added from time $t(i-1)$ to $t(i)$ at the SiC-SiO$_2$ interface, and the outer SiO$_2$ increment added from time $t(0)$ to $t(i-1)$ (Fig. 1). The axial, radial, and hoop stresses in the SiC fiber at time $t(i)$ are $\sigma_z^{SiC}(i)$, $\sigma_r^{SiC}(i)$, and $\sigma_\theta^{SiC}(i)$ respectively, and in the two SiO$_2$ increments are $\sigma_z^{SiO2}(i)$, $\sigma_r^{SiO2}(i)$, $\sigma_\theta^{SiO2}(i)$ and $\sigma_z^{SiO2}(i-1)$, $\sigma_r^{SiO2}(i-1)$, $\sigma_\theta^{SiO2}(i-1)$, respectively. The increment in SiO$_2$ thickness formed from $t=i-1$ to $t=i$ is $w(i)-w(i-1)$. The effect of the ith SiO$_2$ increment on growth stresses is found using strain compatibility equations, where the dilational strain is:[30]

$$\Delta\varepsilon = \sqrt[3]{\frac{1}{\Omega}} - 1 \qquad [6]$$

and

$$\sigma_z^{SiC}(i) = \frac{-E_{SiC}f}{\pi\left(E_{SiC}b^2(i) + E_{SiO_2}\left((b(i)+w(i)-w(i-1))^2 - b^2(i)\right)\right)} + \sigma_z^{SiC}(i-1) \qquad [7]$$

$$\sigma_r^{SiC}(i) = \sigma_r^{SiC}(i-1) - p_{is} \qquad [8]$$

$$\sigma_\theta^{SiC}(i) = \sigma_\theta^{SiC}(i-1) - p_{is} \qquad [9]$$

$$\sigma_z^{SiO_2}(i) = \frac{-E_{SiO_2}f}{\pi\left(E_{SiC}b(i)^2 + E_{SiO_2}\left((b(i)+w(i))^2 - b(i)^2\right)\right)} \qquad [10]$$

$$\sigma_r^{SiO_2}(i) = -p_i \qquad [11]$$

$$\sigma_\theta^{SiO_2}(i) = \frac{p_i(b(i)+w(i))}{w(i)-w(i-1)} \qquad [12]$$

$$\sigma_z^{SiO_2}(i-1) = \frac{f}{\pi\left[(b(i-1)+w(i-1)-w(i))^2 - b^2(i-1)\right]} \qquad [13]$$

$$\sigma_r^{SiO_2}(i-1) = 0 \qquad [14]$$

$$\sigma_\theta^{SiO_2}(i-1) = \frac{2p_{is}b(i-1)^2 - p_i\left((b(i-1)+w(i-1))^2 + b(i-1)^2\right)}{(b(i-1)+w(i-1))^2 - b(i-1)^2} \qquad [15]$$

where **f** is the axial force, **p$_{is}$** is the pressure across the SiO$_2$-SiC interface, **p$_i$** is the pressure across the interface between the ith and the i-1th SiO$_2$ increments, **E$_{SiC}$** and **E$_{SiO2}$** are Young's modulus of the SiC fiber and the SiO$_2$ scale, respectively, and **v$_{SiC}$** and **v$_{SiO2}$** are Poison's ratio for the SiC fiber and the SiO$_2$ scale. Stresses in older increments ($j = i$-2 to $j = 0$) are updated with the stress values in [13 - 15]:

Relaxation of Elastic Stress
    The relaxation of the elastic stresses for all increments ($j=1$ to i) in time increment $\Delta t=t(i)-t(i-1)$ are calculated next. The Eyring model for shear-stress (τ) dependence of glass viscosity (η) is used for SiO$_2$:[11-12,16,31-34]

$$\eta = \eta_o \frac{\tau V_c / 2kT}{\text{Sinh}\left(\tau V_c / 2kT\right)} = \eta_o \frac{\tau / \tau_c}{\text{Sinh}\left(\tau / \tau_c\right)} \qquad [16]$$

V_c is the activation volume for plasticity in SiO$_2$.[11-12,34] k is Boltzmann's constant, τ_c is the critical shear stress above which plasticity is significant (typically ~100 MPa), and η_o is the stress-free SiO$_2$ viscosity:[35-36]

$$\eta_o = C_o \, e^{\frac{Q}{RT}} \qquad [17]$$

where C_o and Q are the pre-exponential factor and activation energy for stress-free viscosity, respectively.[35-36] Shear stress relaxation obeys a Maxwell viscoelastic model:[11,34,37]

$$\frac{d\tau}{dt} = -G \tau / \eta(\tau) \, , \qquad \tau[t(i-1)] = \tau_o \qquad [18]$$

where G is the SiO$_2$ shear modulus, and the initial elastic shear stress at $t(i-1)$ is τ_o. The relaxation of τ_o to a new value (τ) in time increment Δt for all the increments ($j=1$ to i) is determined by substitution of [16-17] in [18] and solving differential equation [18] for τ:

$$\tau = \frac{4kT}{V_c} \text{Coth}^{-1} \left[\frac{e^{\frac{Gt}{\eta_o}}}{\sqrt{\text{Tanh}\left[\frac{V_c \tau_o(j)}{4kT}\right]^2}} \right] = 2\tau_c \text{Coth}^{-1} \left[\frac{e^{\frac{Gt}{\eta_o}}}{\sqrt{\text{Tanh}\left[\frac{\tau_o(j)}{2\tau_c}\right]^2}} \right] \qquad [19]$$

τ_o is determined from the principal stresses for all the increments ($j=1$ to i) by the usual method:[38]

$$\tau_o(j) = \left[\frac{1}{2} s_{ij}(j) s_{ij}(j)\right]^{1/2} \qquad [20]$$

$s_{ij}(j)$ are the components of the deviatoric stress tensor of the jth annular element at time $t(i)$:

$$s_{ij}(j) = \sigma_{ij}(j) - \frac{1}{3}\delta_{ij} \, \sigma_{kk}(j) \qquad [21]$$

Relaxation of $\sigma_\theta(j)$ and $\sigma_z(j)$ is proportional to $\tau(j)/\tau_o(j)$ and to their difference with $\sigma_r(j)$, which is a boundary condition, being zero at the SiO$_2$ surface and near zero elsewhere. The relaxed values of $\sigma_\theta(j)$ and $\sigma_z(j)$ ($\sigma_\theta(j)'$ and $\sigma_z(j)'$) are determined by solution of:

$$\begin{bmatrix} \sigma_\theta(j)' & 0 \\ 0 & \sigma_z(j)' \end{bmatrix} = \frac{\tau(j)}{\tau_o(j)} \begin{bmatrix} \sigma_\theta(j) - \sigma_r(j) & 0 \\ 0 & \sigma_z(j) - \sigma_r(j) \end{bmatrix} + \begin{bmatrix} \sigma_r(j) & 0 \\ 0 & \sigma_r(j) \end{bmatrix} \qquad [22]$$

Radial Displacement and Hoop Stress Generation
Relaxation expands the SiO$_2$ scale radially. The individual radial displacement of the jth increment (u_r) is:

$$u_r(j) = \frac{\Omega_{SiO_2}}{\Omega_{SiC}\left(1 + \varepsilon_z^{SiO2}(j) + \varepsilon_\theta^{SiO2}(j) + \varepsilon_z^{SiO2}(j)\varepsilon_\theta^{SiO2}(j)\right)} - 1 \qquad [23]$$

where

$$\varepsilon_z^{SiO2}(j) = \frac{1}{E_{SiO_2}}\left[\sigma_z^{SiO2}(j) - \upsilon_{SiO_2}\left(\sigma_\theta^{SiO2}(j) + \sigma_r^{SiO2}(j)\right)\right] \qquad [24]$$

$$\varepsilon_\theta^{SiO2}(j) = \frac{1}{E_{SiO_2}}\left[\sigma_\theta^{SiO2}(j) - \nu_{SiO_2}\left(\sigma_z^{SiO2}(j) + \sigma_r^{SiO2}(j)\right)\right] \tag{25}$$

The total radial displacement of the jth increment is the sum of the displacements of younger increments. This adds hoop strain ($\varepsilon_\theta^{SiO2}$) to outer layers as they are forced to a larger circumference (Fig. 1). The new hoop strain in each increment is:

$$\varepsilon_\theta^{SiO2}(j)' = \varepsilon_\theta^{SiO2}(j) + \sum_j^i u_r(j)\frac{b(j-1)-b(j)}{b(j)} \tag{26}$$

Recalculation of Elastic Stress in the Scale and SiC Fiber after Radial Displacement
　　The stresses in each SiO$_2$ increment are recalculated for $\varepsilon_\theta^{SiO2}(j)'$ by solving the three strain compatibility equations for the three principal stresses. Revised axial stress σ_z^{SiC} is computed from the force exerted by the SiO$_2$ scale, which is the sum of the axial stress in each SiO$_2$ increment × area of that increment:

$$\sigma_z^{SiC}(i) = -\sum_{j=1}^{i-1}\frac{\sigma_z^{SiO2}(j)[2c(i)(w(j)-w(j-1)+w(j)^2 - w(j+1)^2]}{b(i)^2} \tag{27}$$

The revised radial and hoop stress in the SiC fiber are calculated by determining the net pressure (p_n) from the sum of the pressures in each annular increment:

$$\sigma_r^{SiC}(i) = \sigma_\theta^{SiC}(i) = -p_n = -\sum_{j=1}^{i-1}\sigma_\theta^{SiO2}(j)\frac{w(j)-w(j-1)}{b(i)} \tag{28}$$

These revised stresses are added to the next increment i as the program loops back to equations [1 -28]. For SiC oxidation when $w(i) \ll b(i)$, the stress in SiC is much smaller than that in SiO$_2$.

RESULTS AND DISCUSSION

General
　　Growth stress calculations were done using [1-28] in a Mathematica^{TM} program using scale discretization to 500 layers (i=500). Calculations were done for oxidation of Hi-Nicalon^{TM}-S SiC fiber for amorphous scales of 10, 100, 300, 1000, and 3000 nm thickness (**w**) at 700, 800, 900, 1000, 1100, 1200, and 1300°C. The Deal-Grove oxidation kinetics for this fiber have been reported for dry air between 700 and 1300°C, with $A_o = 6.5 \times 10^{-4}$ m, $B_o = 1.2 \times 10^{-8}$ m^2/s, $Q_A = 111$ kJ/mol, and $Q_B = 249$ kJ/mol.[1,39] The initial fiber radius (b_0) is 6.1 µm. The molar volumes for SiC (Ω_{SiC}) and amorphous SiO$_2$ (Ω_{SiO2}) are 27.34 cm^3 and 12.46 cm^3, respectively. The Young's modulus (**E**) and Poisson's ratio (**v**) values used for SiC and SiO$_2$ were $E_{SiC} = 400$ GPa, $\nu_{SiC} = 0.157$, $E_{SiO2} = 70$ GPa, and $\nu_{SiO2} = 0.17$. The shear modulus for silica (**G**) is 34 GPa, and has only weak temperature dependence.[40] The pre-exponential factor C_o and activation energy **Q** for stress-free SiO$_2$ viscosity are 3.8×10^{-13} Pa·s and 712 kJ/mol, respectively.[35-36] The activation volume for plasticity in SiO$_2$, V_c, decreases with temperature, and has been inferred to have values ranging from 1.2×10^{-28} m^3 to 3×10^{-28} m^3.[11-12,34] V_c corresponds to a critical shear stress τ_c of about 100 MPa, which is roughly consistent with experiment at 500 to 1400°C.[32,41] The variation in growth stress with change in fiber radius was examined by calculations using $b_o = 2$ µm and "flat-plate" $b_0 \rightarrow \infty$ (1 km). Calculations for σ_θ, σ_z, σ_r, and τ at **T** of 700° - 1300°C for w = 300 n m and $b_o = 2$ µm, 6 µm, and "flat-plates" are shown in figure 2. Calculations for other scale thicknesses and for crystallized scales are reported elsewhere.[22]
　　The accuracy of growth stress calculations is limited by silica viscosity accuracy. Silica scales formed during SiC oxidation incorporate carbon.[42-45] Network carbon in amorphous SiO$_2$ stiffens the network structure, making it more viscous and less permeable to O$_2$.[46-48] However, SiO$_2$ viscosity may also be reduced by incorporation of impurities in Hi-Nicalon^{TM}-S fiber.[39]

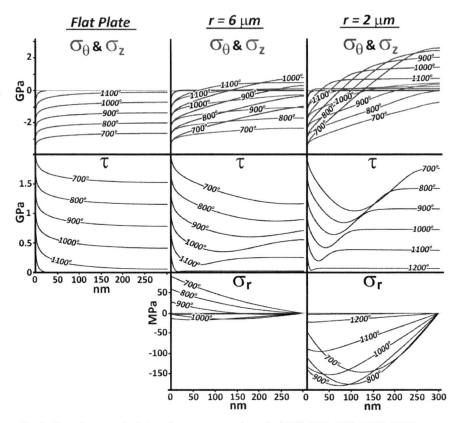

Fig. 2. *Growth stress calculations for σ_θ, σ_z, σ_r, and τ at T of 700°, 800°, 900°, 1000°, 1100°, 1200°, and 1300°C for a 2 µm radius fiber, 6 µm radius fiber, and a flat-plate for a 300 nm amorphous SiO₂ scale on Hi-Nicalon™-S SiC fiber.*

General features for hoop stress (σ_θ) and axial stress (σ_z) in amorphous SiO₂ scale are evident in figure 2 for 300 nm, thick scales on 2 µm and 6 µm SiC fibers and "flat-plate" SiC. The continuous changes of σ_θ and σ_z with change in b_0 at 700°C to 1300°C throughout the scale are evident. Compressive elastic stress of ~-25 GPa for σ_θ and σ_z is rapidly relaxed by the Eyring model [16] for SiO₂ glass viscosity to ~-4 GPa at 700°C and ~-1 GPa at 1200°C just 1 nm away from the SiC-SiO₂ interface. At T >1200°C glass viscosities are so low that stress levels are insignificant more than 20 nm from the SiC-SiO₂ interface. For flat plates $\sigma_\theta = \sigma_z$, and stress is always compressive. σ_z is driven towards tensile values by the Poisson effect from the tensile σ_θ caused by outward radial expansion. Note that a 300 nm scale takes ~10⁸ s to form at 700°C and ~10⁵ s to form at 1000°C, so the latter is of more practical interest. Calculations for crystalline SiO₂ scale are reported elsewhere.[22] The viscosities of crystalline scale are much higher than those for amorphous scale, so growth stresses in crystalline scale are much larger than amorphous scale.

For a 2 µm radius fiber, tensile σ_θ develops at the scale surface at all temperatures, and reaches 3 GPa at 700°C. Tensile σ_z does not develop until T > 800°C and reaches 300 MPa at 1000°C. For a 6 μm radius fiber, tensile σ_θ does not develop until $\mathbf{T} > 850°C$ and reaches a maximum of about 600 MPa at 1000°C; tensile σ_z does not develop until 1100°C and reaches 100 MPa at that temperature. For thicker scales, larger fractions of the scale are under tensile stress.[22]

For a 2 μm radius fiber, compressive σ_r up to -200 MPa is present in the center of the scale at 800° - 900°C. This is a consequence of tensile σ_θ at the scale surface. σ_r decreases towards the SiC-SiO$_2$ interface from the high compressive σ_θ near that interface. For a 6 μm radius fiber σ_r is tensile at 700° – 900°C because σ_θ is compressive at the scale surface and throughout the scale thickness, and only reaches small compressive values at $T > 900°C$ when significant amounts of the scale are in tensile.

τ of ~12 GPa is rapidly relaxed to ~ 2 GPa at 700°C and 500 MPa at 1100°C just 10 nm away from the SiC-SiO$_2$ interface (Fig. 2). It continues to decrease as compressive σ_z and σ_θ decreases with distance from the SiC-SiO$_2$ interface until σ_θ becomes tensile, at which point τ increases towards the SiO$_2$ surface, reaching surface values > 1.5 GPa at 700°C for a 2 μm radius fiber and > 1 GPa for a 6 μm radius fiber, and values close to 500 MPa at 1000°C for both fiber radii..

Steady-State Tensile Stress
For the 300 nm thick scale a "steady-state" develops for σ_θ tensile stress for $T > 700°C$ for $\mathbf{b_o} = $ 2 μm and $T \geq 1000°C$ for $\mathbf{b_o} = 6$ μm. For thicker scales the steady-state region develops at lower temperatures.[22] For $\tau < \tau_c$ of 100 MPa [16] where a stress–free viscosity (η_o) [17] is applicable, an analytical expression for the steady-state hoop stress [$\sigma_\theta(ss)$] can be derived:[22]

$$\sigma_\theta \ (ss) = \frac{2B\left(\frac{1}{\Omega} - 1\right)\eta_o}{wb} \qquad [29]$$

For $\tau > 100$ MPa the stress dependence of viscosity is significant, and η [16] must be substituted for η_o, giving:

$$\sigma_\theta \ (ss) = -2\tau_c \textbf{Csch}\left[\frac{bw\tau_c\Omega}{B\eta_o(\Omega - 1)}\right]^{-1} \qquad [30]$$

Comparison of predictions for steady-state $\sigma_\theta(ss)$ tensile stress from [30] for a 6 μm radius ($\mathbf{b_o}$) fiber with numerical calculations are shown to be very close in another publication.[22]

By insetting the Arrhenius expressions for $\mathbf{B}$ and η_o into [29], the decrease in steady-state tensile stress with increasing temperature is a consequence of $\mathbf{Q} > \mathbf{Q_b}$:

$$\sigma_\theta \ (ss) = \frac{2B_oC_oe^{\frac{Q-Q_b}{RT}}\left(\frac{1}{\Omega} - 1\right)}{wb} \qquad [31]$$

If materials exist for which $\mathbf{Q} < \mathbf{Q_b}$, an increase in tensile growth stress with increase in oxidation temperature is expected.

Summary and Conclusions
A method to calculate the axial, hoop, and radial growth stresses in SiO$_2$ scales generated by the 2.2× volume expansion during SiC fiber oxidation was developed. The method assumes the initial oxidation volume expansion is equal in all directions (dilatational) and that the stresses resulting from constraint of that expansion are relaxed radially with an Eyring stress-dependent SiO$_2$ viscosity, although other appropriate viscosity models can be substituted. The method can be equally well applied to fibers of silicon or other materials. High compressive hoop and axial stresses of ~25 GPa are very quickly relaxed to much lower values at all temperatures. Radial expansion creates tensile hoop stress in the outer scale. Tensile hoop stress eventually drives axial stress to a tensile state by the Poisson effect. Tensile hoop and axial stress can reach values > 2 and 0.5 GPa, respectively for oxidation for long times at 700° - 900°C on Hi-Nicalon[TM]-S fibers. At temperatures greater than 1200°C growth stresses are quickly relaxed to negligible levels by viscous flow of SiO$_2$. The accuracy of the growth stress calculation method is likely to be limited by knowledge of accurate values for amorphous silica viscosity. Tensile hoop stresses reach steady-state values that can be described by

analytical expressions. The decrease in tensile hoop gr owth stress with increase in oxidation temperature is a consequence of activation energy for viscous flow > activation energy for oxidation.

References

1   Hay, R. S. *et al.* Relationships between Fiber Strength, Passive oxidation and Scale Crystallization Kinetics of Hi-Nicalon^{TM}-S SiC Fibers. *Ceram. Eng. Sci. Proc.* **32**, 39-54 (2011).

2   Chollon, G. *et al.* A Model SiC-Based Fibre with a Low Oxygen Content Prepared from a Polycarbosilane Precursor. *J. Mater. Sci.* **32**, 893-911 (1997).

3   Chollon, G. *et al.* Thermal Stability of a PCS-Derived SiC Fibre with a Low Oxygen Content (Hi-Nicalon). *J. Mater. Sci.* **32**, 327-347 (1997).

4   Hsueh, C. H. & Evans, A. G. Oxidation Induced Stresses and Some Effects on the Behavior of Oxide Films. *J. Appl. Phys.* **54**, 6672-6686 (1983).

5   Morscher, G. N. & Pujar, V. V. Design Guidelines for In-Plane Mechanical Properties of SiC Fiber-Reinforced Melt-Infiltrated SiC Composites. *Int. J. Appl. Ceram. Technol.* **6**, 151-163 (2009).

6   Morscher, G. N. Tensile creep and rupture of 2D-woven SiC/SiC composites for high temperature applications. *J. Eur. Ceram. Soc.* **30**, 2209-2221 (2010).

7   Garikipati, K. & Rao, V. S. Recent Advances in Models for Thermal Oxidation of Silicon. *J. Computational Physics* **174**, 138-170 (2001).

8   Rao, V. S. & Hughes, T. J. R. On Modelling Thermal Oxidation of Silicon I: Theory. *Int. J. Numerical Methods in Engineering* **47**, 341-358, doi:10.1002/(sici)1097-0207(20000110/30)47:1/3<341::aid-nme774>3.0.co;2-z (2000).

9   EerNisse, E. P. Stress in Thermal SiO_2 during Growth. *Appl. Phys. Lett.* **35**, 8-10 (1979).

10 Rafferty, C. S. & Dutton, R. W. Plastic Analysis of Cylinder Oxidation. *Appl. Phys. Lett.* **54**, 1815-1817 (1989).

11 Sutardja, P. & Oldham, W. G. Modeling of stress effects in silicon oxidation. *Electron Devices, IEEE Transactions on* **36**, 2415-2421 (1989).

12 Delph, T. J. Intrinsic strain in SiO_2 thin films. *J. Appl. Phys.* **83**, 786-792 (1998).

13 Pomp, A., Zelenka, S., Strecker, N. & Fichtner, W. Viscoelastic Material Behavior: Models and Discretization Used in Process Simulator DIOS. *IEEE Trans. Elec. Dev.* **47**, 1999-2007 (2000).

14 Uematsu, M. *et al.* Two-dimensional simulation of pattern-dependent oxidation of silicon nanostructures on silicon-on-insulator substrates. *Solid-State Electronics* **48**, 1073-1078 (2004).

15 Kao, D.-B., McVittie, J. P., Nix, W. D. & Saraswat, K. C. Two-Dimensional Thermal Oxidation of Silicon - II. Modeling Stress Effects in Wet Oxides. *IEEE Trans. Electron. Dev.* **35**, 25-37 (1988).

16 Rafferty, C. S., Borucki, L. & Dutton, R. W. Plastic Flow During the Thermal Oxidation of Silicon. *Appl. Phys. Lett.* **54**, 1516-1518 (1989).

17 Oh, E., Walton, J., Lagoudas, D. & Slattery, J. Evolution of stresses in a simple class of oxidation problems. *Acta Mechanica* **181**, 231-255 (2006).

18 Mihalyi, A., Jaccodine, R. J. & Delph, T. J. Stress effects in the oxidation of planar silicon substrates. *Appl. Phys. Lett.* **74**, 1981-1983 (1999).

19 Yen, J.-Y. & Hwu, J.-G. Enhancement of silicon oxidation rate due to tensile mechanical stress. *Appl. Phys. Lett.* **76**, 1834-1835 (2000).

20 Yen, J.-Y. & Hwu, J.-G. Stress effect on the kinetics of silicon thermal oxidation. *J. Appl. Phys.* **89**, 3027-3032 (2001).

21 Gauthier, W., Pailler, F., Lamon, J. & Pailler, R. Oxidation of Silicon Carbide Fibers During Static Fatigue in Air at Intermediate Temperatures. *J. Am. Ceram. Soc.* **92**, 2067-2073 (2009).

22 Hay, R. S. Growth Stress in SiO_2 during Oxidation of SiC Fibers. *J. Appl. Phys.* (submitted).

23 Brown, D. K., Hu, S. M. & Morrissey, J. M. Flaws in Sidewall Oxides Grown on Polysilicon Gate. *J. Electrochem. Soc.* **129**, 1084-1089 (1982).

24 Navi, M. & Dunham, S. T. A Viscous Compressible Model for Stress Generation/Relaxation in SiO_2. *J. Electrochem. Soc.* **144**, 367-371 (1997).

25 Hu, S. M. Stress-related Problems in Silicon Technology. *J. Appl. Phys.* **70**, R53-R80 (1991).

26   Causin, P., Restelli, M. & Sacco, R. A Simulation System Based on Mixed-hybrid Finite Elements for Thermal Oxidation in Semiconductor Technology. *Computer Methods in Applied Mechanics and Engineering* **193**, 3687-3710 (2004).
27   Senez, V., Collard, D., Ferreira, P. & Baccus, B. Two-dimensional Simulation of Local Oxidation of Silicon: Calibrated Viscoelastic Flow Analysis. *IEEE Trans. Elec. Dev.* **43**, 720-731 (1996).
28   Wilson, L. O. & Marcus, R. B. Oxidation of Curved Silicon Surfaces. *J. Electrochem. Soc.* **134**, 481-490 (1987).
29   Deal, B. E. & Grove, A. S. General Relationships for the Thermal Oxidation of Silicon. *J. Appl. Phys.* **36**, 3770-3778 (1965).
30   Tsui, Y. C. & Clyne, T. W. An Analytical Model for Predicting Residual Stresses in Progressively Deposited Coatings Part 2: Cylindrical Geometry. *Thin Solid Films* **306**, 34-51 (1997).
31   Eyring, H. Viscosity, Plasticity, and Diffusion as Examples of Absolute Reaction Rates. *J. Chem. Phys.* **4**, 283-291 (1936).
32   Donnadieu, P. P., Jaoul, O. & Kleman, M. Plasticit de la silice amorphe de part et d'autre de la transition vitreuse *Philos. Mag. A* **52**, 5-17 (1985).
33   Uchida, T. & Nishi, K. Formulation of a Viscoelastic Stress Problem Using Analytical Integration and Its Application to Visocelastic Oxidation Simulation. *Jap. J. Appl. Phys.* **40**, 6711-6719 (2001).
34   Senez, V., Collard, D., Baccus, B., Brault, M. & Lebailley, J. Analysis and Application of a Viscoelastic Model for Silicon Oxidation. *J. Appl. Phys.* **76**, 3285-3295 (1994).
35   Doremus, R. H. Viscosity of Silica. *J. Appl. Phys.* **92**, 7619-7629 (2002).
36   Hetherington, G., Jack, K. H. & Kennedy, J. C. Viscosity of Vitreous Silica. *Phys. Chem. Glasses* **5**, 130-136 (1964).
37   Malvern, L. E. *Introduction to the Mechanics of a Continuous Medium*. 1st edn, 713 (Prentice-Hall, 1969).
38   Frost, H. J. & Ashby, M. F. *Deformation Mechanism Maps*. (Pergamon Press, 1982).
39   Hay, R. S. *et al.* Hi-Nicalon^{TM}-S SiC Fiber Oxidation and Scale Crystallization Kinetics. *J. Am. Ceram. Soc.* **94**, 3983-3991 (2011).
40   Polian, A. & et al. Elastic properties of a-SiO$_2$ up to 2300 K from Brillouin scattering measurements. *Europhys. Lett.* **57**, 375 (2002).
41   Li, J. H. & Uhlmann, D. R. The flow of glass at high stress levels: I. Non-Newtonian behavior of homogeneous 0.08 Rb$_2$O·0.92 SiO$_2$ glasses. *J. Non. Cryst. Sol.* **3**, 127-147 (1970).
42   Narushima, T., Kato, M., Murase, S., Ouchi, C. & Iguchi, Y. Oxidation of Silicon and Silicon Carbide in Ozone-Containing Atmospheres at 973K. *J. Am. Ceram. Soc.* **85**, 2049-2055 (2002).
43   Chaudhry, M. I. A Study of Native Oxides of b-SiC Using Auger Electron Spectroscopy. *J. Mater. Res.* **4**, 404-407 (1989).
44   Ramberg, C. E., Cruciani, G., Spear, K. E., Tressler, R. E. & Ramberg, C. F. Passive-Oxidation Kinetics of High-Purity Silicon Carbide from 800 to 1100 C. *J. Am. Ceram. Soc.* **79**, 2897-2911 (1996).
45   Ogbuji, U. J. T. & Opila, E. J. A Comparison of the Oxidation Kinetics of SiC and Si$_3$N$_4$. *J. Electrochem. Soc.* **142**, 925-930 (1995).
46   Renlund, G. M., Prochazka, S. & Doremus, R. H. Silicon Oxycarbide Glasses: Part II. Structure and Properties. *J. Mater. Res.* **6**, 2723-2734 (1991).
47   Rouxel, T., Massouras, G. & Soraru, G.-D. High Temperature Behavior of a Gel-Derived SiOC Glass: Elasticity and Viscosity. *J. Sol-Gel Sci. Tech.* **14**, 87-94 (1999).
48   Rouxel, T., Soraru, G.-D. & Vicens, J. Creep Viscosity and Stress Relaxation of Gel-Derived Silicon Oxycarbide Glasses. *J. Am. Ceram. Soc.* **84**, 1052-1058 (2001).

Advanced Materials and Processing for Photonics and Energy

EFFECT OF CHROMIUM-DOPING ON THE CRYSTALLIZATION AND PHASE STABILITY IN ANODIZED TiO$_2$ NANOTUBES

I.M. Low[1], H. Albetran[1], V. De La Prida[2], P. Manurung[3] and M. Ionescu[4]
[1]Centre for Materials Research, Curtin University, GPO Box U1987, Perth, WA 6845, Australia
[2]Department of Physics, University of Oviedo, Spain
[3]Department of Physics, University of Lampung, Indonesia
[4]Australian Nuclear Science and Technology Organisation, Sydney, NSW 2234, Australia

ABSTRACT
    Production of limitless hydrogen fuel by visible light splitting of water using the photo-electrochemical technology is cost-effective and sustainable. To make this an attractive viable technology will require the design of TiO$_2$ photocatalyst capable of harnessing the energy of visible light. One possible solution is the doping of TiO$_2$ to reduce its band gap. In this paper, the effect of Cr-doping by ion-implantation on the crystallisation and phase stability of TiO$_2$ nanotubes at elevated temperature is described. The effect of Cr-doping on the resultant microstructures, phase changes and composition depth profiles are discussed in terms of synchrotron radiation diffraction, scanning electron microscopy, and ion-beam analysis by Rutherford backscattering spectrometry.

INTRODUCTION
    Titanium dioxide (TiO$_2$) is a wide band-gap semiconductor with energy of 3.0-3.2 eV. It is widely used in applications such as hydrogen production, gas sensors, photocatalytic activities, dye-sensitized solar cells and photo-electrochemical cells because of its relative high efficiency and high stability. However, due to its wide band gap energy, TiO$_2$ is active only under near-ultraviolet irradiation. Therefore, numerous studies have been carried out over the last 20 years to develop modified TiO$_2$ so that they are active under visible light irradiation (> 400 nm). One of the most studied methods is by doping the TiO$_2$ materials with metal ions (iron, nickel, cobalt, vanadium, and chromium)[1-4] or non-metallic elements (nitrogen, sulphur and carbon).[5-14] For instance, Cr or N-doping of TiO$_2$ can result in visible light response by virtue of a narrowing in the band gap due to the mixing of p states of nitrogen with O 2p states. Among these doping methods, doping with transition metals is one of the most efficient methods. X-ray photoelectron spectroscopy (XPS) results showed that the doped samples contained N, C, B and F elements and the doped TiO$_2$ showed the shift in the band gap transition down to 2.98 eV. The photocatalytic activity of the doped TiO$_2$ was 1.61 times better than undoped TiO$_2$. Researchers have conducted doping of bismuth into TiO$_2$ to enhance the photocatalytic activities in these systems. It has been reported that the metal/TiO$_2$ nanostructures enhance the efficiency of photocatalysis in water-splitting and dye-sensitized solar cell (DSSC). Metals compounded on semiconductor materials increase charge-collection efficiency due to a much slower electron-hole recombination, giving rise to longer electron lifetime, which will result in an increasing interfacial electron-transfer process.
    Hitherto, the doping of TiO$_2$ has been widely synthesized using the sol-gel method.[15-19] However, ion-implantation has now emerged as an alternative but effective doping method to improve the separation of the photo-generated electron-hole pairs or to extend the wavelength range of the TiO$_2$ photo-responses into the visible region.[20-25] XPS measurements revealed that the implanted nitrogen species were mainly interstitial ones. The nitrogen concentration was increased with increasing ion flux which could be controlled by adjusting the gas flow rate of the ion source, resulting in improved visible-light photocatalytic activities. Higher visible-light photocatalytic efficiency was achieved with higher implanted nitrogen concentration.

EXPERIMENTAL METHODS

Sample Preparation

Ti foils (99.6 % purity) with dimensions of $10 \times 10 \times 0.1$ mm^3 were used for the anodizing to produce self-organized and well-aligned TiO$_2$ nano-tube arrays. The process of potentiostatic anodization was performed in a standard two-electrode electrochemical cell, with Ti as the working electrode and platinum as the counter electrode. Prior to anodization, Ti-foils were degreased by sonicating in ethanol, isopropanol and acetone for 5 minutes each, followed by rinsing with deionised water, and then drying using nitrogen stream. After drying, the foils were exposed to the electrolyte which consists of 100 ml of Ethylene glycol + 2.04 ml of water + 0.34 g of NH$_4$F. The electrolyte's pH was kept constant at pH = 6, and its temperature was kept at about 25 $^\circ$C. The anodization process was performed under an applied voltage of 60 V for 20 h.[26] After that, the resulting TiO$_2$ structures were then rinsed in ethanol, immersed in hexamethyl-disilazane (HMDS) and dried in air.

Doping by Ion-Implantation

The as-anodized samples were doped with Cr ions using a MEVVA Ion Implanter with a dose of $\sim 7 \times 10^{14}$ ions/cm^2 that corresponds to 8,500 pulses. The near surface composition depth profiling of ion-implanted samples was measured by ion-beam analysis and Rutherford backscattering spectrometry (RBS) using He^{1+} ions at 1.8 MeV. This phase of work was conducted at the Australian Nuclear Science and Technological Organisation (ANSTO). The information obtained allowed the calculation of depth distribution of implanted species and the implanted dose.

The morphologies of the anodized and ion-implanted samples were characterized using a focussed ion-beam scanning electron microscope (FIB-SEM) operating at working distances of 5 mm with an accelerating voltage of 5 kV.

In Situ High-Temperature Synchrotron Radiation Diffraction

In this study, the in-situ crystallisation behaviour of as-anodized samples with or without Cr-doping was characterised using high-temperature synchrotron radiation diffraction (SRD) up to 900 $^\circ$C in argon. All measurements were conducted at the Australian Synchrotron using the Powder Diffraction beamline in conjunction with an Anton Parr HTK20 furnace and the Mythen II microstrip detector. The SRD data were collected at an incident angle of 3° and wavelength of 0.11267 nm.

The phase transitions or structural changes were simultaneously recorded as SRD patterns. These patterns were acquired at high temperatures over an angular range of 100° in 2θ. The SRD patterns were acquired in steps of 100 $^\circ$C from 100 $^\circ$C to 900 $^\circ$C. The collected SRD data were analysed using the CMPR program to evaluate the integrated peak intensities of all phases present. The sum of all the integrated peak intensities in 2θ range of 10-50° was used to calculate the relative phase content of all the phases present at each temperature as follows:

$$W_A = \left(\frac{I_A}{I_A + I_R + I_T} \right) \times 100 \tag{1}$$

where W_A is the wt% of anatase, I_A is sum integrated intensity of anatase, I_R is sum integrated intensity of rutile and I_T is sum integrated intensity of titanium. A similar ratio method was used to calculate the wt% of rutile (W_R) or titanium (I_T). This method is expected to be more accurate than the method of Spurr & Myers[27] where only the peak intensities of anatase (101) and rutile (110) are used. The use of integrated peak intensities in equation (1) can minimise issues relating to preferred orientation, grain size and/or degree of crystallinity.

The mean crystallite sizes (L) of anatase and rutile were calculated from (101) and (110) reflections using the Scherrer equation:[28]

$$L = \frac{K\lambda}{\beta \cos \theta} \qquad (2)$$

where K is the shape factor, λ is the x-ray wavelength, β is the line broadening at half the maximum intensity (FWHM) in radians, and θ is the Bragg angle.

RESULTS AND DISCUSSION
In-Situ Formation of TiO_2 at Elevated Temperature
    Figure 1 shows the synchrotron radiation diffraction plots of Cr-doped TiO_2 nanotubes before and after thermal annealing in air at 20 °C – 900 °C. The TiO_2 nanotube arrays were initially amorphous but eventually crystallized into anatase at 300 °C. The peak intensities of anatase increased rapidly from 300° to 400°C and continued to increase until 900°C. The formation of rutile commenced at 600°C but its growth was sluggish where the peak intensities increased very slightly from 600 to 800°C. Surprisingly, rutile disappeared at 900C with the concomitant appearance of TiO_{2-x} where x = 0.2. This suggests that rutile was unstable in argon atmosphere at 900 °C and decomposed to form a non-stoichiometric titanium oxide and oxygen vacancies (O_v) as follows:

$$5TiO_2 \xrightarrow{900° C / \mathrm{argon}} Ti_5O_9 + O_v \qquad (3)$$

    In contrast, no decomposition of rutile was observed in un-doped TiO_2 nanotubes (see Fig. 2). When compared to Cr-doped TiO_2, both anatase and rutile in the un-doped sample crystallized at a higher temperature, i.e. 400 and 700°C respectively. The corresponding phase abundances as a function of temperature are shown in Figure 3. The fraction of anatase-to-rutile transformation in the samples calcined at between 600 – 1000 °C can be calculated from the followed equation:[27]

$$x = 1/[1 + 0.8(I_A / I_R]$$

where I_A is the X-ray integrated intensities of the (101) reflection of anatase at ~18.3°, while I_R is that of the (110) reflection of rutile at ~19.8°; x is the weight fraction of rutile in the nanotubes, which is calculated for the Cr-doped sample to be 2.9 wt% for the nanotubes annealed at 600°C, 4.9 wt% at 700°C, and 4.6 wt% at 800°C. The corresponding x values for the un-doped sample were 3.3 wt% at 700°C, 19.0 wt% at 800°C and 15.9 wt% at 900°C.
    It appears that the presence of Cr-doping has accelerated the formation of anatase and rutile at a lower temperature, probably through the facilitation of defects such as vacancies. It is widely accepted that the onset temperature of the anatase to rutile phase transformation and the rate at which it proceeds can be affected significantly by dopants, firing atmosphere, microstructure, sample morphology, and the presence of impurities in the material.[29] The process of the anatase to rutile transformation is known to occur via (a) the nucleation of rutile at point defects, oxygen vacancies, secondary phase inclusions, particle surfaces, and/or at (112) twin interfaces in anatase, and (b) the subsequent consumption of the anatase phase by the growing rutile phase. The formation of rutile from the metastable phase anatase is reconstructive and so takes place through atomic rearrangement involving the breaking of two of the six Ti-O bonds in the TiO_6 octahedra.[30] As anatase transforms to rutile, significant grain growth takes place, resulting in lower surface area and thus poorer photocatalytic performance.
    Transition metals (e.g. Cr) of variable valence are reported to enter the titania lattice and create oxygen vacancies through reduction effects as follows:[29,31]

$$2M^{n+} + O^{2-} \rightarrow 2M^{(n-1)+} + \frac{1}{2}O_2 + O_v \qquad\qquad (4)$$

where M denotes a transition metal atom and O_v denotes an oxygen vacancy. Once substituted for Ti^{4+}, the reduction of such species gives rise to new oxygen vacancies and so enhances rutile formation from metastable anatase through easing of structural rearrangement. These oxygen vacancies are believed to be responsible for crystallisation of anatase at a lower temperature in Cr-doped TiO$_2$ as well as destabilisation and subsequent decomposition of rutile to form non-stoichiometric TiO$_{2-x}$.

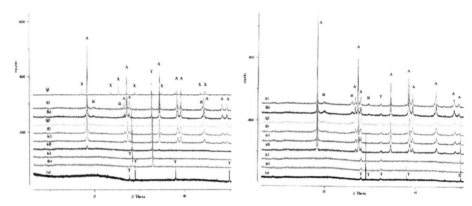

Fig. 1: In-situ synchrotron radiation diffraction plots of Cr-doped TiO$_2$ showing the existence of anatase (A), rutile (R), Ti$_3$O$_9$ (X) and titanium (T) in the temperature range 20 – 900 °C. [Legend: a = 20°C, b = 100°C, c = 200°C, d = 300°C, e = 400°C, f = 500°C, g = 600°C, h = 700°C, i = 800°C, j = 900°C]

Fig. 2: In-situ synchrotron radiation diffraction plots of un-doped TiO$_2$ showing the existence of anatase (A), rutile (R), and titanium (T) in the temperature range 100 – 900 °C. [Legend: (a) = 100°C, (b) = 200°C, (c) = 300°C, (d) = 400°C, (e) = 500°C, (f) = 600°C, (g) = 700°C, (h) = 800°C, (i) = 900°C]

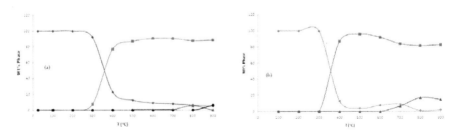

Fig. 3: Variations of phase abundances as a function of temperature for (a) Cr-doped TiO$_2$ and (b) un-doped TiO$_2$. [Legend: titanium (♦); anatase (■); rutile (▲); TiO$_{2-x}$ (●)]

Although anatase remained stable up to 800-900 °C, there was a distinct narrowing and sharpening in the (101) peak, resulting in a corresponding decrease in the values of full-width half-maximum (FWHM) as the temperature increased. A similar sharpening of the (110) peak was observed for rutile. From the Scherrer equation,[28] a decrease in the value of FWHM implies an increase in the mean crystallite size for anatase or rutile. Figure 4 shows the influence of annealing temperature on crystallite size for anatase and rutile in both undoped and Cr-doped TiO$_2$. The crystallite size of anatase was just over 50 nm at 400 °C for un-doped TiO$_2$ but just under 50 nm in Cr-doped TiO$_2$ and increased gradually with temperature to 85 nm and 100 nm respectively at 900°C. In both samples, the crystallite size of rutile was much smaller than that of anatase, especially in the un-doped sample.

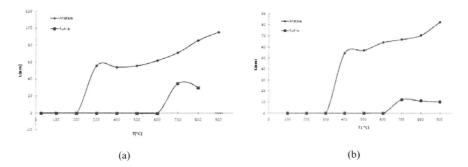

(a) (b)

Fig. 4: Variations of crystallite size of anatase (♦) and rutile (■) as a function of temperature for (a) Cr-doped TiO$_2$ and (b) un-doped TiO$_2$.

Microstructures of TiO$_2$ and Cr-Doped TiO$_2$

After anodizing in the electrolyte of NH$_4$F / ethylene glycol at 25 °C for 20 h, vertically oriented and highly ordered arrays of TiO$_2$ nanotubes with diameter of ~80 nm formed on the Ti-foil.[26] Fig. 5(a) shows the FESEM image of the amorphous TiO$_2$ nanotubes. Upon annealing at 900 °C for 1 h, some anatase transformed into rutile, which appeared as elongated grains randomly distributed amongst the nanotubes shown in Fig. 5 (b). The microstructure of Cr-doped TiO$_2$ is shown in Fig. 5 (c) where the existence of implanted Cr ions has been verified by the EDS analysis of a bright region in Spectrum 1 (see Fig. 5d).

(a)

(b)

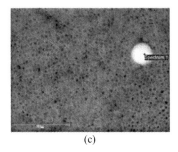
(c)

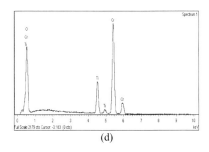
(d)

Fig. 5: Scanning electron micrographs showing the microstructure of (a) as-anodized TiO$_2$, (b) annealed TiO$_2$ at 900°C, (c) Cr-doped TiO$_2$ and (d) EDS analysis of composition in Spectrum 1.

Rutherford Backscattering Spectroscopy (RBS)
    The RBS results in Figure 6 show the concentration of elements (at%) versus depth (at/cm^2) for both the un-doped and Cr-doped samples. In Fig. 6a, we show the depth profile of Ti and O for as-anodized TiO$_2$. The composition at the near surface of TiO$_2$ layer is fairly constant but a composition gradation in both O and Ti exists at the interface between the substrate and the oxide layer. In the doped sample (Fig. 6b), a maximum Cr concentration of 0.7 at% has been implanted in TiO$_2$ down to a depth of about 300 mono layers. A composition gradation also exists at the interface between the substrate and the oxide layer.

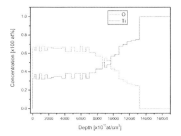
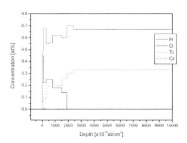

Fig. 6: Composition depth profiles in (a) un-doped TiO$_2$ and (b) Cr-doped TiO$_2$.

CONCLUSIONS
    High-temperature synchrotron radiation diffraction has been used to study the effect of Cr-doping on in-situ crystallization of anatase and rutile in as-anodized TiO$_2$ nanotubes. The as-anodized TiO$_2$ was amorphous but crystallized into anatase and rutile at 400 °C and 700°C in the un-doped sample, but 300°C and 600°C in the doped sample. The rutile formed in the un-doped sample was very stable in argon after annealing at 900°C but was unstable in the doped sample and decomposed to form non-stoichiometric TiO$_{2-x}$ at 900°C. In both cases, increasing the temperature from 300 to 900 °C caused the FWHM of anatase (101) and rutile (110) peaks to decrease, resulting in a concomitant coarsening in crystallite size. The surface of annealed sample exhibited well-aligned, uniform TiO$_2$ nanotubes with

an average diameter of ~80 nm and wall thickness of ~10 nm. Ion-beam analysis by RBS has confirmed the existence of Cr ions in the doped sample and composition gradation within the annealed TiO$_2$ sample at substrate-oxide interface.

ACKNOWLEDGMENTS

This work was supported by funding from the Australian Synchrotron (PD-3611), the Australian Institute of Nuclear Science and Engineering (AINGRA-11134) and Spanish MAT2010-20798-C05-04. We thank Ms E. Miller of Curtin Centre for Materials Research for assistance with the SEM work. The scientific support from the University of Oviedo SCT's, particularly to the Nanoporous Membranes Unit, is also acknowledged.

REFERENCES
[1]M. Anpo, Y. Ichihashi, M. Takeuchi and H. Yamashita, *Res. Chem. Intermed.* **24**, 143 (1998).
[2]A. Zaleska, J.W. Sobczak, E. Grabowska and J. Hupka, Preparation and photocatalytic activity of boron-modifled TiO$_2$ under UV and visible light. *Appl. Catal.* B **78**, 92-100 (2008).
[3]G.B. Saupe, T.U. Zhao, J. Bang, N.R. Desu, G.A. Carballo, R. Ordonem and T. Bubphamala, Evaluation of a new porous titanium-niobium mixed oxide for photocatalytic water decontamination. *Microchemical J.* **81**, 156-162 (2005).
[4]D.B. Hamal and K.J. Klabunde, Synthesis, characterization, and visible light activity of new nanoparticle photocatalysts based on silver, carbon, and sulfur-doped TiO$_2$. *J. Colloid & Interfa. Sci.* **311**, 514-522 (2007).
[5]J. Virkutyte, B. Baruwati and R.S. Varma, Visible light induced photobleaching of methylene blue over melamine-doped TiO$_2$ nanocatalyst. *Nanoscale* 2, 1109-1111 (2010).
[6]T. Umebayashi, T. Yamaki and S. Tanaka, Visible light-induced degradation of methylene blue on S-doped TiO$_2$. *Chem. Lett.* **32**, 330-331 (2003).
[7]S. Sakthivel, M. Janczarek and H. Kisck, Visible light activity and photo-electrochemical properties of nitrogen-doped TiO$_2$. *J. Phys. Chem.* B **108**, 19384-19387 (2004).
[8]Y. Choi, T. Umebayashi and M. Yoshikawa, Fabrication and characterization of C-doped anatase TiO$_2$ photocatalysts. *J. Mater. Sci.* **39**, 1837-1839 (2004).
[9]H. Irie, Y. Watanabe and K. Hashimoto, Carbon-doped anatase TiO$_2$ powders as a visible-light sensitive photocatalyst. *Chem. Lett.* **32**, 772-773 (2003).
[10]S. Sakthivel and H. Kisch, Daylight photocatalysts by carbon-modified titanium dioxide. *Angew. Chem. Int. Edn.*, **42**, 4908-4911 (2003).
[11]R. Asahi, T. Morikawa and T. Ohwaki, Visible-light photocatalysis in nitrogen-doped titanium oxides. *Science* **293**, 269-273 (2001).
[12]W. Ren, Z. Ai, F. Jia, L. Zhang, X. Fan and Z. Zou, Low temperature preparation and visible light photocatalytic activity of mesoporous carbon-doped crystalline TiO$_2$. *Appl. Catal.* B **69**, 138-144 (2007).
[13]T. Ohno, M. Akiyoshi, T. Umebayashi, K. Asai, T. Mitsui and M. Matsumura, Preparation of S-doped TiO$_2$ photocatalysts and their photocatalytic activities under visible light. *Appl. Catal.* A **265**, 115-121 (2004).
[14]C. Lettmann, K. Hildebrand, H. Kisch, W. Macyk and W.F. Maier, Visible light photodegradation of 4-chlorophenol with a coke-containing titanium dioxide photocatalyst. *Appl. Catal.* B **32**, 215-227 (2001).
[15]Y. Kuroda, T. Mori, K. Yagi, N. Makihata, Y. Kawahara, M. Nagao and S. Kittaka, Preparation of visible-light-responsive TiO$_{2-x}$N$_x$ photocatalyst by a sol-gel method: Analysis of the active center on TiO$_2$ that reacts with NH$_3$. *Langmuir* **21**, 8026-8034 (2005).

[16]A. Hattori, M. Yamamoto, H. Tada and S. Ito, A promoting effect of NH$_4$F addition on the photocatalytic activity of sol-gel TiO$_2$ films. *Chem. Lett.* **27**, 707-708 (1998).

[17]Y. Shen, T. Xiong, T. Li and K. Yang, Tungsten and nitrogen co-doped TiO$_2$ nano-powders with strong visible light response. *Appl. Cataly. B: Environ.* **83**, 177-185 (2008).

[18]T.C. Jagadale, S.P. Takale, R.S. Sonawane, H.M. Joshi, S.I. Patil, B.B. Kale and S.B. Ogale, N-doped TiO$_2$ nanoparticle based visible light photocatalyst by modified peroxide sol-gel method. *J. Phys. Chem. C* **112**, 14595-14602 (2008).

[19]Y. Tseng, C. Kuo, C. Huang, Y. Li, P. Chou, C. Cheng and M. Wong, Visible-light-responsive nano-TiO$_2$ with mixed crystal lattice and its photocatalytic activity. *Nanotechno.* **17**, 2490-2497 (2006).

[20]T. Yamaki, T. Sumita and S. Yamamoto, Formation of TiO$_{2-x}$F$_x$ compounds in fluorine-implanted TiO$_2$. *J. Mater. Sci. Lett.* **21**, 33-35 (2002).

[21]H. Shen, L. Mi, P. Xu, W. Shen and P.N. Wang, Visible-light photocatalysis of nitrogen-doped TiO$_2$ nanoparticulate films prepared by low-energy ion implantation. *Appl. Surf. Sci.* **253**, 7024-7028 (2007).

[22]J. Park, J.Y. Lee and J.H. Cho, Ultraviolet-visible absorption spectra of N-doped TiO$_2$ film deposited on sapphire. *J. Appl. Phys.* **100**, 113534 (2006).

[23]H. Yamashita, Y. Ichihashi and M. Takeuchi, Characterization of metal ion-implanted titanium oxide photocatalysts operating under visible light irradiation. *J. Synchrotron Radia.* **6**, 451-452 (1999).

[24]T. Yamaki, T. Umebayashi and T. Sumita, Fluorine-doping in titanium dioxide by ion implantation technique. *Nucl. Instr. & Meth. Phys. Res. B* **206**, 254-258 (2003).

[25]H. Yamashita, M. Harada, J. Misaka, M. Takeushi and M. Anpo, Degradation of propanol diluted in water under visible light irradiation using metal ion-implanted titanium dioxide photocatalysts. *J. Photochem. Photobiol. A* **148**, 257-261 (2002).

[26]V. Vega, M.A. Cerdeira, V.M. Prida, D. Alberts, N. Bordel, R. Pereiro, F. Mera, S. García, M. Hernández-Vélez and M. Vázquez, Electrolyte influence on the anodic synthesis of TiO$_2$ nanotube arrays. *J. Non-Cryst. Solids* **354**, 5233-5235 (2008).

[27]R.A. Spurr and H. Myers, Quantitative analysis of anatase-rutile mixtures with an x-ray diffractometer. *Anal. Chem.* **29**, 760–762 (1957).

[28]B.D. Cullity and S.R. Stock, *Elements of X-Ray Diffraction*, 3rd Ed., Prentice-Hall Inc., pp. 167-171 (2001).

[29]D. Hanaor and C. Sorrell, Review of the anatase to rutile phase transformation. *J. Mater. Sci.* **46**, 1-20 (2011).

[30]S. Riyas, G. Krishnan and P.N. Mohandas, Anatase-rutile transformation in doped titania under argon and hydrogen atmospheres. *Adv. Appl. Ceram.* **106**, 255 – 26 (2007).

[31]K.J.D. Mackenzie, Calcination of titania V. Kinetics and mechanism of the anatase-rutile transformation in the presence of additives. *Trans. J. Bri. Ceram. Soc.* **74**, 77-84 (1975).

FRONTIERS IN NANOMATERIALS AND NANOTECHNOLOGY AND IMPACT ON SOCIETY

J. Narayan
Department of Materials Science and Engineering, EB I, Suite 3030, Centennial Campus
North Carolina State University, Raleigh, NC 27695-7907, USA.

ABSTRACT

Throughout the human history, major changes in our civilization and quality of life have occurred as a result of discoveries of new materials. From Stone Age to Iron, to Bronze, to Semiconductors and Nanomaterials, materials have revolutionized and defined our civilization. Semiconductors ushered in the modern era of electronics and information age. Nanomaterials stand to revolutionize the modern era as we must do more with less and less to conserve and sustain our dwindling resources of critical materials. For every technology, there is a "materials bottleneck" and this aspect is amplified considerably for nanotechnology. This paper focuses primarily on the impact of nanomaterials and nanotechnology on the conservation of materials resources and the sustenance of welfare of the society. The special emphasis is on the discoveries of novel materials and processing, and the transition of nanomaterials to nanotechnology to manufacturing for the good of the society, specifically related to nanostructuring of materials for next-generation systems having superior performance. We start with a discussion of intrinsic advantages of nanoscale materials and systematic approach for transition into systems. As the feature (grain) size of solid-state materials decreases, the defect content reduces and below a critical size material can be defect-free. Since these critical sizes for most materials lie in 5-100nm, there is a fundamental advantage and an unprecedented opportunity to realize the property of a perfect material. Along with this opportunity, there is a major challenge with respect to large fraction of atoms at the interfaces, which must be engineered to realize the advantages of nanotechnology based systems. We specifically address nanosystems based upon nanodots and nanolayered materials synthesized by thin film deposition techniques, where recurring themes include nanostructuring of materials to improve performance; thin film epitaxy across the misfit scale for orientation controls; control of defects, interfaces and strains; and integration of nanoscale devices with (100) silicon based microelectronics and nanoelectronics. The systems of interest are based upon strong novel structural materials, nanomagnetics for information storage, nanostructured or Nano Pocket LEDs, variety of smart structures based upon vanadium oxide and novel perovskites integrated with Si(100), and nanotechnology based solutions to enhance fuel efficiency and reduce environmental pollution.

Keywords: Nanotechnology, manufacturing, epitaxial self-assembly, domain matching epitaxy, epitaxy across the misfit scale, nanomagnetics, Nano Pocket Light Emitting Diodes, Vanadium Oxide based Smart Structures

I.   INTRODUCTION

With his legendary quote, "There is a plenty of room at the bottom," Feynman (1959) envisioned profound impact of nanotechnology in our lives (1). In 1983, Narayan et al. showed very interesting and useful mechanical and optical property modifications of ceramics such as magnesium oxide by incorporating metallic colloids (nanoparticles) in the matrix (2). This vision was reinforced by Smalley (2004), who predicted future global energy prosperity through nanotechnology by reducing the consumption of useful materials (3). Since nanomaterials hold the key to the success of nanotechnology by assuming the most important component by being at the bottom of the nanotechnology food chain, it is imperative that we examine the fundamental advantages of nanoscale materials. We have shown that there is a critical size below which material can be free of defects (4-5). Of course, this critical size is a strong function of the nature of defects, for example, vacancies, interstitials, dislocations, twins, and dislocations. It is interesting to note that these critical sizes for most materials lie in the nanometer range (4-5). Thus, nanotechnology affords the opportunity to create and realize the property of a perfect material. However, along with this opportunity comes a big challenge, as the fraction of atoms on surfaces (Ω) increases inversely with grain size (d) as $\Omega = 6\delta/d$. Assuming δ as 1nm, we have number of atoms on surfaces equal to that in the bulk, when d is 6nm. The bonding characteristics of these surface atoms need to be controlled to achieve the desirable properties in nanosystems. Thus there are tremendous challenges related to synthesis and processing (3-D self-assembly) interface engineering, atomic-scale characterization and modeling (6-12). These challenges can be addressed by utilizing the transition paradigm, shown in the form of an octahedron, where nanoscience (nanomaterials) constitute the base with components of synthesis and processing, nanoscale characterization, structure-property correlations and nanoscale modeling. These components need to be carried out synergistically before proceeding to nanosystems and eventually transition to manufacturing.   This review covers the following topics systematically: (a) synthesis and processing of nanodots and nanolayered structures by controlling thin film growth modes; (b) control of orientation of nanostructures through the paradigm of domain matching epitaxy (13-14); (c) nanomagnetics and information storage over Terabits per chip (6); (d) solid state lighting based upon III-nitride and II-oxide film film heterostructures with a focus on thin film epitaxy and Nano Pocket LEDs (15-18); (e) nanostructured VO_2 and perovskite based smart sensors(19-22); and (f) nanotechnology solutions to enhance fuel efficiency and reduce pollution (19). The recurring themes in all of these topics include nanostructuring of materials to improve performance; thin film epitaxy across the misfit scale; control of stresses/strains, defects and interfaces; and integration of device functionality specifically with Si(100) based microelectronics/nanoelectronics.

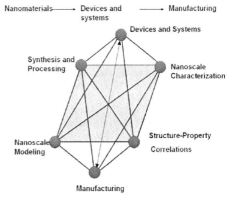

Fig. 1 Nanoscience to Nanotechnology to Manufacturing Octahedron

II. THIN FILM GROWTH MODES FOR PROCESSING NANOSTRUCTURED MATERIALS

Thin film growth modes can be used in a controlled way to produce self-assembled nanodots or nanolayered structures (5-8). As illustrated in Figure 2, there are three distinct modes of thin film growth: two-dimensional (layer by layer growth; Frank- Vander Merwe growth); three dimensional (island growth, Volmer-Weber growth); and a mixture of 2-D and 3-D growth (Stranski-Krastnov). These growth modes depend upon surface free energy of the substrate (σ_s), surface free energy of the film (σ_f), and the interfacial free energy (σ_{sf}). The two-dimensional layer-by-layer growth occurs when $\sigma_s > \sigma_f + \sigma_{sf}$, where σ_s is high and σ_s and σ_{sf} are low. This two-dimensional growth mode can be used to create layered nanostructures. The three-dimensional growth mode occurs when $\sigma_s < \sigma_f + \sigma_{sf}$, where σ_s is low and σ_s and σ_{sf} are high.

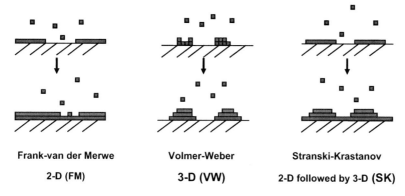

Frank-van der Merwe   **Volmer-Weber**   **Stranski-Krastanov**

2-D (FM)   **3-D (VW)**   **2-D followed by 3-D (SK)**

Fig. 2 Thin film growth modes for nanolayered and self-assembled nanodot structures

During thin film deposition using pulsed lasers (pulsed laser deposition), the islands or nanodots of uniform size can be created by controlling the laser deposition and substrate variables.

Figure 3 shows a schematic of a pulsed laser deposition system, where high-power Excimer laser is used to ablate material in a controlled way with a precision of a fraction of a monolayer. The typical laser parameters are: pulse energy density 3-5 Jcm^{-2}, pulse duration 25ns, and laser wavelength with photon energy 4-6eV. The large laser power in the range of 120-600mW leads to nonequilibrium stoichiometric evaporation, where the average energy of evaporated species ranges from 100-1000kT. The laser deposition variables include pulse energy density, pulse duration, laser wavelength and repetition rate. The substrate variables include substrate temperature and surface free energy, which can be varied by

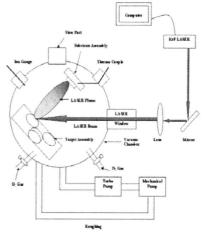

Fig. 3 Pulsed laser deposition system with nanosecond Excimer lasers

introducing appropriate surfactant interlayers.

To produce nanostructured materials consisting of islands, we lay down the first layer of islands by controlling the kinetics of clustering, where nanoclusters are self organized as result of a kinetically driven free energy minimum. The total energy of the system has four components: volume free energy, surface free energy, interface free energy, and interaction free energy. The total free energy has two components: nanodot size dependent and nanodot size independent terms. The total energy of the system shows a local minimum in free energy which can be used to obtain self-organized nanodots (6). The idea is to create self organization driven kinetically by the local minimum, while avoiding the global thermodynamic minimum which leads to Ostwald ripening. After the formation of the first layer of nanodots, an interfacial layer is introduced to start the renucleation process for the second set of nanodots, as shown in Fig. 4. The characteristics of strains and interfacial layers are critical for renucleation and self-organization of the second layer. Fig 4(a) shows a cross-section TEM with a corresponding schematic in Fig. 4(b), and magnetic properties of these self-assembled structures are shown in Fig. 4(c).

III.   DOMAIN MATCHING EPITAXY

In the following, we introduce a new paradigm for thin film growth to deal with epitaxial growth across the misfit scale (13-14). In the domain matching epitaxy, we consider the matching of lattice planes, which could be different in different directions of the film-substrate interface. The misfit is accommodated by matching of integral multiples of lattice planes, and there is one extra half-plane (dislocation) corresponding to each domain.

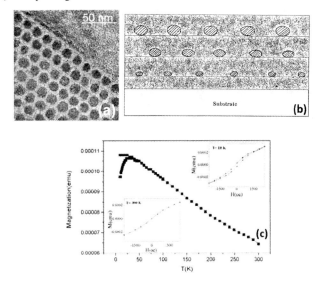

Fig. 4 Self-assembled nanostructured materials: a) TEM X-section; (b) Schematic; and (c) Magnetic Properties

The misfit can range from being very small to very large. In the small misfit regime, the DME reduces to LME where matching of the same planes or lattice constants is considered with a misfit typically less that 7-8%. If the misfit falls in between the perfect matching ratios of planes, then the size of the domain can vary in a systematic way to accommodate the additional misfit. In the conventional lattice matching epitaxy (LME), the initial or unrelaxed misfit strain is (ϵ_c) is given by $\epsilon_c = a_f / a_s - 1$, where a_f and a_s are lattice constant of the film and the substrate, respectively. In LME, the ϵ_c is less than 7 to 8% which is relaxed by the introduction of dislocations beyond the critical thickness during thin film growth. In the domain matching epitaxy, the matching of lattice planes of the film (d_f) with the those of the substrate (d_s) is considered with similar crystal symmetry. In DME, the film and the substrate planes could be quite different as long as they maintain the crystal symmetry. The LME, on the other hand, involves the matching of the same planes between the film and the substrate. In DME, the initial misfit strain ($\epsilon = d_f / d_s - 1$) could be very large, but this can be relaxed by matching of m planes of the film with n of the substrate. This matching of integral multiples of lattice planes leaves a residual strain of ϵ_r given by

$$\epsilon_r = (md_f / nd_s - 1) \text{-------------(1)},$$

where m and n are simple integers. In the case of a perfect matching $md_f = nd_s$, and the residual strain ϵ_r is zero. If ϵ_r is finite, then two domains may alternate with a certain frequency to provide for a perfect matching according to,

$$(m+\alpha)d_f = (n+\alpha)d_s \text{----------------(2)},$$

where α is the frequency factor, for example, if $\alpha = 0.5$, then m/n and (m+1)/ (n+1) domains alternate with an equal frequency.

Assuming $d_f > ds$, we have n>m.

Therefore,

$$n - m = 1 \text{ or } f(m) \text{-----------(3)}.$$

The difference between n and m could be 1 or some function of m. In Fig. 1, n -m = 1 forϵ = 0 –50% and n – m = f(m) for ϵ = 50 to 100%.

From equations (1) through (3), we can derive,

$$(m + \alpha) = 1 \text{ or } f(m) \text{----------------(4)}.$$

The equation (4) basically governs the domain epitaxy, as plotted in Fig. 5. Fig. 5 shows a general plot of misfit percent strain as a function of ratio of film/substrate lattice constants of major planes matching across the interface.

Table I provides a summary of different systems which have been grown with various misfit strains. The table also includes the systems which fall in between the two

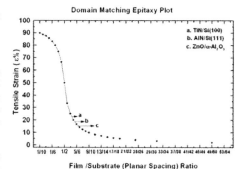

Fig. 5 Domain matching epitaxy plot of strain versus film-substrate planar spacing ratio

domains where two domains alternate with periodicity needed for a complete relaxation. It should be noted that a 45^0 rotation in some cubic systems and a 30^0 rotation in certain hexagonal systems are part of the domain matching concept involving the matching of major planes between the film and the substrate. The plot in Fig. 5 provides a unified framework of lattice matching and domain matching epitaxy with misfit strain ranging from 2-90% (50% corresponding to 1/2 matching).

If the domain matching is not perfect, epitaxy occurs by accommodating the additional misfit by changing the domain size, controlled by the parameter ⊠. In this framework, it is important to realize that the nature of dislocations remains the same, only their periodicity changes. These points will become clearer as we discuss specific cases of domain matching epitaxy and the nature of dislocation, including their periodicity.

Table I: Domain Epitaxy systems

m/n	Planar spacing ratio	Experimental examples	Strain ε%
1/10	0.1		90.0%
1/9	0.11		88.8%
1/8	0.125		87.5%
1/7	0.143		85.7%
1/6	0.166		83.3%
1/5	0.20		80.0%
1/4	0.25		75.0%
1/3	0.33		66.7%
1/2 and 1/3	0.33–0.50	Mo, Nb, Ta, W/Si; Ni₃Al/Si(100)	50.0%
1/2	0.50	Fe/Si, Cr/Si, NiAl//Si(100)	
2/3	0.666	Cu/Si(100)	33.33%
3/4	0.750	TiN/Si(100)	25.00%
4/5	0.80	AlN/Si(111)	20.00%
5/6	0.83	α–Al₂O₃/ZnO(0001)	16.67%
6/7	0.857	α–Al₂O₃/ZnO(0001), Cu/TiN(100)	14.29%
7/8	0.8750	α–Al₂O₃/GaN(0001)	12.50%
8/9	0.888	α–Al₂O₃/AlN(0001),	11.11%
9/10	0.90	YBa₂Cu₃O₇ δ /MgO(001)	10.0%
11/12	0.9166	YBa₂Cu₃O₇ δ /MgO(001)	8.33%
12/13	0.9230	STO/MgO(001)	7.69%
13/14	0.3286		7.14%
14/15	0.9333		6.67%
16/17	0.9412		5.88%
17/18	0.9444		5.55%
18/19	0.9474		5.26%
19/20	0.9500		5.00%
20/21	0.9524		4.76%
22/23	0.9556		4.35%
24/25	0.96	Ge/Si(100)	4.0%
31/32	0.9687		3.13%
49/50	0.98	Ge–Si/Si(100)	2.0%

IV. TiN/Si(100) SYSTEM

Epitaxial growth of TiN on silicon substrate represents a major milestone for next-generation semiconductor devices for direct ohmic contacts as well as for discussion barriers in copper metallization. However, with a misfit of over 22% for cube-on-cube TiN (a=0.424 nm) epitaxy over silicon (a=0.543 nm), it is beyond the critical strain (7-8%) of conventional lattice matching. However, epitaxial growth of TiN on silicon substrate was demonstrated by the concept of domain matching epitaxy. The films were grown using the standard pulsed laser deposition method described in the previous section. Fig. 6 shows a detailed high-resolution cross-section TEM micrograph, where the 3/4 and 4/5 domains alternate. The micrograph was taken in the <110> zone axis of Si and TiN, it is interesting to note the matching of {111} extra half planes in silicon as well as TiN. The implications of these on the nature of dislocations will be discussed later in the section. From Fig. 6, the lattice misfit of 22% lies in the middle 3/4 and 4/5 matching, which explains the alternating of domains. In fact with $\alpha = 0.5$ (equation 2), $3.5a_{Si}=19.01$ matches quite well with $4.5 \times a_{TiN} = 19.08$, which also represents the size of domain for the system with virtually no residual misfit. In our earlier study, we considered various energy terms for epitaxial growth of TiN on Si(100) and found a significant reduction in energy due to domain matching epitaxy compared to the unrelaxed state (10).

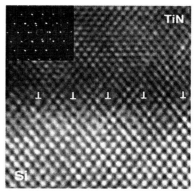

Fig. 6 Domain matching epitaxy of TiN on Si(100) with lattice misfit of 22.5%

The nature of dislocations can be established directly from the high-resolution TEM micrographs. The Burgers vector of the dislocations is determined to be a/2 <110> lying in {111} planes. The two sets of a/2 <110> dislocations combine at the interface to produce a/2 <110> dislocations lying in the {001} interface. This dislocation reaction can be described as: a/2[101](11-1) + a/2[01-1](111) → a/2[110](001). In some cases, the dislocations do not combine and create an extended core structure associated with the pair of dislocations. The formation of a/2<110> dislocations in {111} plane in TiN with a sodium chloride structure represents a significant finding. The TiN having a sodium chloride structure has {110} slip planes with a/2 <110> Burgers vectors. Only under certain extreme nonequilibrium conditions such as high field, a/2<110> lying in {001} planes have been observed (24). However, this is first for a/2<110> dislocation in {111} planes of sodium chloride structure. These new dislocations or slip systems may impact mechanical and physical properties of TiN films or materials of sodium chloride structure, in general, in a significant way. According to von Mises criterion, five independent slip systems are needed for a crystal to undergo a plastic deformation by slip. In TiN having a sodium chloride structure, there are only two independent a/2<110>{110} slip systems available, which restricts a general deformation, resulting in twinning and fracture. However, with a/2<110>{110}slip systems, there are 384 ways of choosing five independent slip systems, which can lead to a general deformation of TiN (25).

V. FORMATION OF SELF-ASSEMBLED NANOMAGNETIC STRUCTURES

Storing of information in magnetic nanodots and reading them reliably are major challenges in next-generation information storage technology. This challenge requires that nanodots of magnetic materials such as Ni, Ni-Pt, Co, Fe-Pt be grown epitaxially on nonmagnetic substrates. Furthermore, if these nanodots can be integrated epitaxially on Si(100), this platform opens possibilities of next-generation devices with integrated functionalities(26-28). Fig. 7 shows the size as a function of Blocking temperature T_B , below which nanodot will be ferromagnetic and above which the dot will be thermally unstable and it will turn superparamagnetic. This plot is derived from K V(anisotropy energy times volume of the nanodot) = 25 kT_B, where k is the Boltzmann constant. For super high-density information storage device above room temperature, the Fe-Pt is clearly the material of choice.

First challenge for Fe-Pt based nanostructure is to stabilize it in the ordered $L1_0$ FCT (face-centered tetragonal) structure with c/a ratio of 0.97, where Fe and Pt atoms are layered, as shown in Fig. 8. The disordered phase with a random distribution of Fe and Pt atoms, the Fe-Pt has FCC (face-centered cubic) structure with low anisotropy energy of 6.0×10^4 ergs cm^{-3}, compared to 7.0×10^7 ergs cm^{-3} for the ordered FCT structure. Using the paradigm of domain matching epitaxy, we are able to grow Fe-Pt/TiN/Si(100) nanolayered as well as nanodot epitaxial structures, where lattice misfit ranges from 9.5% between Fe-Pt and TiN to 22.5% between TiN and Si(100).

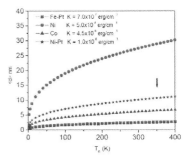

Fig.7 Critical size for ferromagnetism as a function of blocking temperature

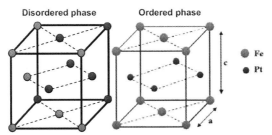

Fig. 8 Disordered and ordered $L1_0$ structure of FePt

The results are shown in Fig. 9, where Fig 9(a) shows Fe-Pt/TiN/Si(100) nanolayered structure in cross-section with corresponding selected-area diffraction (Fig. 9(b)) and high-resolution TEM image (Fig. 9(c)). The alignment of diffraction spots in Fig 9(b) demonstrate epitaxial alignment of Fe-Pt and TiN with Si(100) despite a large lattice misfit. Fig. 10 shows a comparison of magnetic properties between nanolayered and nanodot magnetic structures. When the applied field is perpendicular to the substrate, the coercivity for nanolayered structure is 3,000Oe compared to 10,000 Oe for the case of nanodot structure. These values are quite adequate for information storage applications.

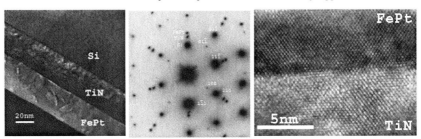

Fig. 9 Cross-section TEM micrographs of $L1_0$ ordered FePt

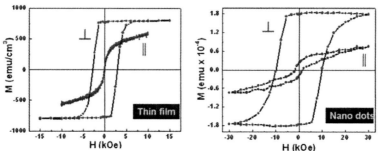

Fig. 10 Magnetization M versus H curves: (a) Nanolayered thin films; and (b) Nanodots

VI.   CRITICAL ISSUES IN HETEROEPITAXY OF LEDS FOR SOLID-STATE LIGHTING

Solid state lighting (SSL) holds revolutionary promise for solving the energy problem and protecting the environment (29-30). As stated in the recent DOE SSL workshop report (31), "SSL still faces enormous technical challenges that are likely to be surmounted only with dramatic increases in scientific understanding." Today's best devices approach ~35% internal quantum efficiency (IQE) on polar and ~50% efficiency on nonpolar substrates, dropping to less than 25% at high powers needed for lighting. Also, there is significant drop in quantum efficiency with increasing wavelength. Thus, blue LEDs are considerably more efficient than green LEDs. The details of these phenomena must be understood to achieve near 100% internal quantum efficiency. The GaN based LEDs fabricated on polar (0001) sapphire have a – ABAB-- stacking of wurtzite structure, which creates a large sheet of charges at the interfaces due to instantaneous polarization along the c-direction. In addition, unrelaxed lattice, thermal and misfit strains produce piezoelectric polarization between adjacent layers. These spontaneous and piezoelectric strain fields reduce the overlap between electron and hole wave functions and reduce luminous efficiency of LEDs. In a typical LED, GaN barriers surround the GaInN quantum wells, where there is a negative charge on n-type side of the wall. This charge repels electrons and hampers charge injection. On the p-side of the wall, there is a positive charge which attracts electrons and increases the probability of electron escape, thus reducing radiative electron-hole recombination. In view of these results, considerable improvements in efficiency are expected for LEDs on nonpolar substrates, if problems related to thin film epitaxy across the misfit scale on polar substrates, control of stresses and strains, structure and properties of quantum wells and barrier layers, and formation of low-resistance Ohmic contacts to top p-GaN layers can be addressed adequately.

The critical challenges for high-efficiency LEDs pertain to III-nitride epitaxy on nonpolar substrates such as r-plane of sapphire and (001) Si, where one has to deal with epitaxy across the misfit scale (13-14). There is an additional complication for the growth of III-nitrides on these nonpolar substrates, because the misfit along the two principal in-plane directions can vary greatly from 1.19% to 16.08% for GaN growth on r-plane of sapphire. In the case of GaN epitaxy on Si (001), the misfit changes its sign from positive to negative along the two <100> directions. However, thin film epitaxy across the misfit scale can be handled by the paradigm of domain matching epitaxy, where integral multiples of lattice planes match across the film-substrate interface (13-14). Since the critical thickness under a large misfit (>10%) is about a monolayer, misfit dislocations can be nucleated at the surface steps from the beginning so that rest of the film can grow strain-free. The relaxation of misfit strain is critical to minimize piezoelectric effects which can hamper radiative carrier recombination and reduce IQE. In order to minimize the number density of threading dislocations, it is critical that initial stages of growth be two-dimensional, so that misfit dislocations spread across the interface and stay confined into the interface without generating threading dislocations.

The GaN growth on sapphire is inherently three-dimensional because $\sigma_s < \sigma_f + \sigma_{sf}$, where σ_s (substrate surface energy) and σ_f (film surface energy) are low, and σ_{sf} (interfacial energy) is high. The GaN islands grow on sapphire substrate and often rotate to accommodate the large lattice misfit. If this rotation exceeds about 1^0, subboundaries are formed leading to formation of high density (over $10^{10}/cm^2$) of dislocations with concomitant decrease in EQE (external quantum efficiency). In addition, hexagonal pits are generated during growth, where LEDs cannot be fabricated. It is found that low-temperature ($\sim 550^0C$) growth leads to formation of pseudo two-dimensional growth with fairly smooth surface. However OMCVD of GaN around 550^0C results in inefficient cracking of NH_3 and nitrogen deficiency. This nitrogen deficiency promotes the formation of cubic GaN (c-GaN) with high-density of stacking faults. The c-GaN transforms into h-GaN during high-temperature growth, but some of the Shockley partials associated with stacking faults eventually lead to formation of threading dislocations (32). Thus there is a challenge to grow stable h-GaN at low-temperatures by two-dimensional growth by OMCVD. Narayan's group has shown the growth of h-GaN by PLD at lower temperatures. It is also envisioned that laser-assisted OMCVD would lead to stabilization of h-GaN at lower temperatures. Two-dimensional growth can also be further induced by a suitable buffer layer.

The second critical issue relates to management of stresses and strains, which is critical as unrelaxed strains will lead to generation of undesirable electric fields through piezoelectric effects. Under the DME paradigm, large lattice misfit strains can be relaxed more easily within the critical thickness of one to two monolayers. The lattice relaxation within two monolayers has been demonstrated though elegant "in-situ" ZnO thin film growth on (0001) sapphire using synchrotron experiments, thus confirming critical aspects of DME paradigm (9-10). In the case of a-GaN on r-plane of sapphire, there is added challenge as the misfit varies from 1.19 % to 16.08%. In the case of a-GaN on (001) Si, the inplane misfit changes sign from positive to negative, which can be engineered to minimize the residual misfit during initial stages of growth so that the rest of the film can grow stress-free. Since the in-plane strains (ϵ_{xx} and ϵ_{yy}) are additive and relate to normal strain (ϵ_{zz}) through Poisson's ratio(v) via: $\epsilon_{zz}/(\epsilon_{xx} + \epsilon_{yy}) = v/(1-v)$, it is possible to manipulate in-plane strains and minimize adverse effects of the normal strain.

The third issue is related to thickness (15-16) and phase separation of quantum wells (34-35). Previous studies by Morkoc group (36) have shown that by reducing the InGaN barrier thickness from 12nm to 3nm, the onset of EQE droop was extended from 200 to 1100 Acm^{-2} for LEDs on polar (0001) sapphire. It is interesting to note that Mahajan's group (34-35) showed that there is a critical thickness, which is around 3nm, above which phase separation in InGaN layer occurs. Narayan et al showed that below 3nm the thickness variation in GaInN can be introduced to confine carriers and enhance EQE, forming thebasis for Kopin's "Nano Pocket" LEDs (15-16). Recently, Schubert's group has shown

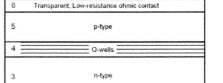

Critical Issues in heteroepitaxy of LEDs

6	Transparent, Low-resistance ohmic contact			
5	p-type			
4	Q-wells			
3	n-type			
2	Nucleation Layer			
1	Substrate : Film			

α-Al$_2$O$_3$(c-plane): Growth of Polar Films : Isotropic biaxial strain
α-Al$_2$O$_3$(r-plane): Growth of Nonpolar Films · Anisotropic biaxial strain

Fig. 11 Schematic cross-section of different layers in a typical LED structure

that LEDs with GaInN quantum wells and polarization-matched AlGaInN barriers exhibit reduced efficiency droop and improved EQE at larger currents compared to conventional GaInN/GaN LEDs on polar (0001) sapphire by reducing electron leakage from the active regions of LEDs (37). Thus by nanostructuring quantum wells, it is possible to reduce electron leakage and minimize nonradiative recombination (15-16).

There is more potential for large energy savings from lighting than from any other area of energy usage. While solid-state lighting in principle can reach efficiencies of 100%, today's best devices reach only ~50% efficiency, dropping to ~30% efficiency at higher power needed for lighting and longer light wavelength. Fig. 11 shows critical issues involved in the improvement of internal quantum efficiencies. These include: (1) formation of nucleation layer on polar (c- sapphire substrate) and nonpoar (r-plane sapphire substrate) via domain matching epitaxy; (2) n-type GaN epitaxial layer; (3)

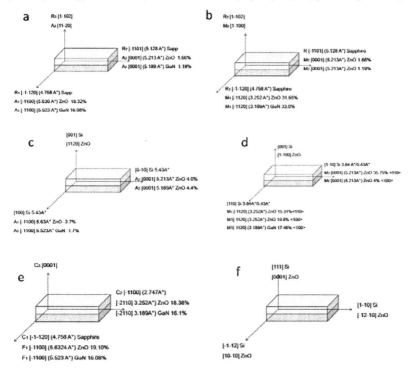

Fig. 12 Epitaxial relations and strains associated with various orientations of GaN and ZnO films on polar and nonpolar silicon and sapphire substrates

formation of quantum wells such as GaN/InGaN; (4) p-type layer GaN layer; and (5) low-resistance transparent ohmic contacts. First four address photon generation and the last one addresses photon extraction to optimize overall quantum efficiency.In this review, we focus on epitaxial growth of III-nitrides and II-oxides on polar and nonpolar silicon and sapphire substrates. The growth on Silicon substrates is critical from cost reduction considerations. Fig. 12 shows epitaxial relationships for polar (GaN and ZnO) a- and m-oriented films on r-plane sapphire and Si(100), and nonpolar c-oriented films on c-plane sapphire and Si(111) substrates. The figure also provides a summary of lattice misfits involved in polar as well nonpolar substrates. The lattice misfit on polar substrates is anisotropic, where the misfit is isotropic on polar substrates. The misfit for a-GaN on r-plane sapphire ranges from -1.19% for [0001] orientation to -16.08% for [-1100] orientation, and corresponding values for ZnO are similar -1.5% and -18.3%, respectively. The misfit for m-GaN on r-plane sapphire ranges from -1.19% for [0001] orientation to 33.0% for [-1120] orientation, and corresponding values for ZnO are similar -1.66% and 31.65%, respectively. The misfit for a-GaN on Si (100) ranges from 4.4% for [0001] orientation to -1.7% for [-1100] orientation, and corresponding values for a-ZnO are similar 4.0% and -3.7%, respectively. The misfit for m-GaN on Si (100) ranges from 4.4% for [0001] orientation to 17.46% for [-1120] orientation, and corresponding values for a-ZnO are similar 4.0% and 19.8%, respectively. On polar c-sapphire and Si(111), the misfit c-GaN and c-ZnO range from 15-18%. Such large lattice misfits can be handled by the paradigm of domain matching epitaxy, as shown for a-ZnO/r-sapphire, c-ZnO/c-sapphire, and c-AlN/Si(100) heterostructures.

VIIa. NONPOLAR A-ZnO ON R-PLANE SAPPHIRE

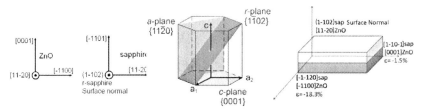

Fig. 13 Schematic of nonpolar a-plane ZnO growth on r-plane sapphire

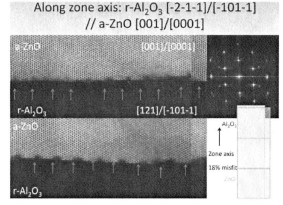

Fig 14. Atomic resolution TEM micrograph with zone axis r-sapphire [-101-1]// a-ZnO[0001], showing misfit dislocations as predicted by DME paradigm

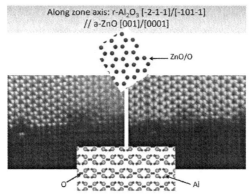

Fig 15. Atomic resolution TEM micrograph with zone axis r-sapphire [-101-1]// a-ZnO[0001] showing the details of core structure of misfit dislocations

Nonpolar a-ZnO can be grown epitaxially by the paradigm of domain matching epitaxy. Fig. 13 shows a-plane (11-20) and r-plane (1-102) in the hexagonal unit cell, both of these planes have rectangular symmetry. The epitaxial alignments of a-ZnO film on r-sapphire substrates are shown separately as well as together, where there is anisotropic misfit of -1.5% along [0001] direction and -18.3% along [-1100] direction. It should be noted that the total misfit consists of three components: (a) lattice misfit; (b) thermal misfit due to differences in the coefficients of thermal expansion; and (c) microstructural misfit due to residual defects.

Detailed x-ray diffraction studies (summarized in Table II) show almost a complete strain relaxation along the large misfit direction [-1100], except for a residual thermal and defect strain.

Table II: In-plane strain in nonpolar a-plane ZnO grown on r-plane Sapphire

	Calculated				Experimental (Measured)	
	Lattice parameter (Å)	Lattice Misfit Strain ε_l (%)	α (E-6K^{-1}) at 873K	Thermal Misfit Strain ε_t(%)	Lattice Parameter (Å)	Residual Strain in Film ε (%)
[-1100]ZnO	5.6288	-18.3	8.47	0.01	5.6386	0.17
[-1-120]sap	4.758		8.25			
[0001]ZnO	5.2066	-1.5	4.936	-0.18	5.186	-0.4
[-1101]sap	15.384		8.59			
[11-20]ZnO	3.2498				3.251	0.04

However, the residual strain along the [0001] direction is higher because smaller strains are more difficult to relax. For smaller strains, critical thickness is large and dislocations have to nucleate on the surface and glide to the interface for strain relaxation. The nucleation barrier and propagation of dislocation steps are more difficult in oxides and nitrides due to their higher atomic bonding energies. Fig. 14 shows atomic-resolution TEM cross-section along the [0001] ZnO zone axis with [-1100] ZnO (large lattice misfit -18.3%) along the interface.

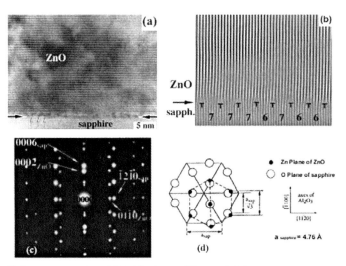

Fig 16: Domain matching of (2-1-10) planes of ZnO with (30-30) planes of sapphire

Here the misfit dislocations are present as dictated by the domain matching epitaxy paradigm. Further details of atomic structure associated with cores of dislocations are shown in Fig. 15. The simulation shows the details atomic arrangements derived from these images.

VIIb. POLAR c-ZnO GROWTH ON c-PLANE SAPPHIRE

Growth (0001) ZnO films on (0001) sapphire occurs by domain matching epitaxy, where integral multiples of film and substrate planes match to accommodate a large lattice misfit. The ZnO has Wurtzite structure with a = 0.3252nm and c = 0.5213nm; and sapphire has corundum structure with a = 0.4758nm and c = 1.299nm. During thin film epitaxy, there is 30^0 or 90^0 rotation in the basal plane to align (2-1-10) planes of ZnO with (30-30) planes of sapphire, which is dictated by the chemical free energy associated with oxygen bonding. To accommodate the planar misfit close to 16%, 5/6 and 6/7 domains alternate with an equal frequency.

The results of the domain epitaxy are shown in Fig. 16, where Fig (a) shows a cross-section image of high-resolution TEM micrograph, Fig (b) shows matching of integral multiples of planes and dislocations associated with the domains, Fig (c) shows relevant epitaxial relationships after the rotation, and Fig (d) is an schematic of ZnO and sapphire unit cells after 30^0 or 90^0 rotation.

VIIc. POLAR c-AlN GROWTH ON Si(111)

Growth of III-nitrides and II-oxides on silicon substrates is critical to reduce the manufacturing costs associated with LEDs for solid state lighting as well as for integrating device functionalities. In the following, we discuss the details of c-AlN growth on Si(111). The basic ideas discussed here are applicable to other silicon systems. Growth of (0001) AlN films on Si(111) substrate occurs by domain matching epitaxy, where integral multiples of film and substrate planes match to accommodate 19% lattice misfit. AlN has a Wurtzite structure with a = 0.3112nm and c = 0.4892nm; and silicon has diamond cubic structure with a = 0.543nm. During thin film epitaxy, there is no 30^0 or 90^0 rotation in the basal plane and (2-1-10) planes of AlN align with (220) planes of silicon.

To accommodate the planar misfit close to 19%, three domains of 5/4 alternate with one 7/6 domain. The results of the domain epitaxy are shown in Fig. 17, where Fig (a) shows a cross-section image of high-resolution TEM micrograph, where planar matching of domains and dislocations associated with these domains are clearly depicted, and Fig (b) shows a schematic of planar alignment in the (111) Si plane.

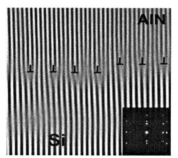 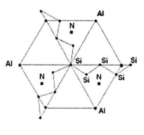

Fig. 17 Domain matching of (2-1-10) planes of ZnO with (220) planes of silicon

VIID. NANOSTRUCTURING OF QUANTUM WELLS TO ENHANCE LED EFFICIENCY

The LEDs based upon three nitrides provide an interesting puzzle regarding their high internal quantum efficiency despite the presence of high (10^9 to 10^{10}/cm^2) dislocation densities in III-nitride/c-sapphire thin film heterostructures. In the conventional III-nitrides, dislocation densities exceeding 10^3 to 10^4/cm^2 are quite detrimental to solid state devices. In the past, it was conjunctured that dislocations in these nitrides do not provide effective traps and recombination centers for active carriers. Some researchers argued for the In segregation and phase separation which can result in bandgap fluctuations leading to quantum confinements of carriers. These confined carriers do not see the traps and recombination centers associated with dislocations in thin film heterostructures. Results from Narayan's group showed the role of thickness variation in quantum confinement without In segregation and phase separation in InGaN/GaN quantum wells(15-16). Mahajan's group performed careful In segregation and phase separation studies as a function InGaN quantum well thickness and showed that there is a critical thickness above which In segregation and phase separation occur (34-35). Recent results from Humphrey's group have shown that In segregation and phase separation in GaN/InGaN quantum wells can occur under certain conditions during electron irradiation of TEM studies (38). Thus there is consensus that the key to enhanced internal quantum efficiency in GaN/InGaN quantum wells is the quantum confinement by thickness variation as proposed and patented by Narayan et al. who coined the word Nano Pocket LEDs (15-16).

Fig 18 shows high-resolution STEM-Z cross-section image of quantum wells: (A) nonuniform quantum wells; and (B) relatively uniform quantum wells. The enhanced contrast, which varies as Z^2 is associated with higher Z (atomic number) of In = 49 compared to Z = 31 for Ga. The output power is enhanced by a factor of 2 in nonuniform quantum wells compared to uniform wells. The thickness variation can be introduced during growth below a critical thickness to minimize the free-energy of the system. Thus, the thickness variation based quantum confinement of carriers is critical to present and next-generation III-nitride based LEDs and lasers.

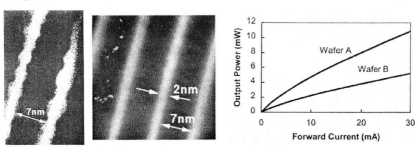

Fig. 18 GaN/GaInN based nanostructured LEDs: (a) Quantum wells with thickness variation (Nano Pocket LEDs); (b) Uniform quantum GaInN/GaN quantum wells; and (c) Internal quantum efficiency comparison.

VIII.   VANADIUM OXIDE BASED INTEGRATED SMART SENSORS

The VO_2 based materials have been used for applications related to infrared sensors. However, there is challenge to enhance the functionality of these devices by integrating them with silicon microelectronics/nanoelectronics. Such devices can be labeled as "smart", because these devices can perform sensing, manipulation and response functions on a single chip.

Vanadium oxide (VO_2) bulk single-crystals exhibit an ultrafast semiconductor to metal transition (SMT) as the crystal structure changes from monoclinic to tetragonal or rutile upon heating close to 68^0C. The transition involves a small lattice distortion along the c-axis, which results in failure of bulk single-crystals, when subjected to repeated heating and cooling cycles. Therefore, thin films, which are able to withstand these distortions and dissipate heat through the substrate, are critical to a variety of technological applications. However, the characteristics of the SMT (sharpness, ΔT, over which electrical and optical property change is completed, amplitude (ΔA) and width of hysteresis (ΔH)) are a strong function of defect microstructure in thin films(19-22), including point defects and clusters, defect chemistry, grain size, and characteristics of grain boundaries. Until now, fundamental correlations, which can relate structure, chemistry and properties, have not been established and a good predictive model does not exist. Unavailability of well-characterized, stoichiometric single-crystal films as well as of polycrystalline (with controlled grain-boundary structure) and amorphous films has been a major impediment to comprehensive structure, chemistry and property correlations. We have grown textured and epitaxial VO_2 films with increasing grain size all the way to high-quality single-crystal on (0001) and (11-20) sapphire substrates by using pulsed laser deposition. High-quality single crystal VO_2 films were grown via domain matching epitaxy, involving matching of integral multiples of lattice planes between the film of monoclinic structure and the sapphire substrate. To integrate with silicon (100), we have utilized a new platform where tetragonal YSZ (yttria stabilized zirconia) is grown on epitaxially on Si(100), which acts as template for the growth of epitaxial VO_2(16). Fig. 19

shows Θ-2Θ (thin film normal) and φ (in-pane) scans from x-ray diffraction. From these results the following epitaxial relationship is established: $VO_2[100]//YSZ[110]//Si[100]$; and $VO_2[010]//YSZ[001]//Si[001]$.

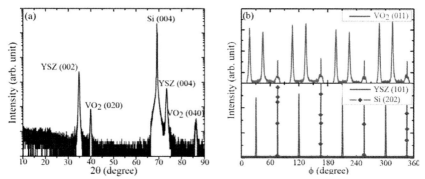

Fig. 19 Θ-2Θ (thin film normal) and φ (in-pane) scans from x-ray diffraction

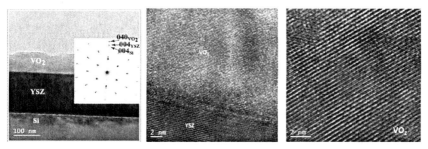

Fig. 20 Cross-section TEM micrographs: (a) VO2/TiN/Si(100) in <011> silicon zone with inset selected area diffraction; (b) and (c) high resolution TEM micrographs in cross section

Fig 20 shows cross-section TEM showing nanolayered structures of VO_2 and YSZ of high-crystalline quality and atomically sharp interfaces. Fig 21 shows semiconductor to metal transition which is quite sharp with ΔA over three orders of magnitude, and hysteresis (ΔH) of 6K. Since c-axis of the VO_2 film is under tensile strain, the transition temperature shifts to higher values, as expected. The sharpness and amplitude of the transition, and the hysteresis upon heating and cooling are found to be a strong function of crystal structure, chemistry and microstructure (grain size, characteristics of grain boundaries, and defect content). Recently it was possible to impart ferromagnetism, yet functionality, in VO_2 films by controlling the intrinsic defect concentration.

Fig 22 shows magnetization of VO_2 films when they were processed under the oxygen partial pressure of 10^{-2} torr. The hysteresis values of 150 and 40 Oe were measured at 10K and 300K, respectively. The effective magnetic moment was determined to be 1.25 and $0.99\mu_B$ at 10^{-3} and 10^{-2} torr of oxygen partial pressure respectively. The blocking temperature was estimated to be around 500K, thus making these devices functional at room temperature quite efficiently.

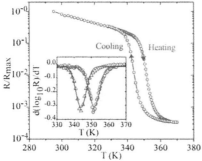

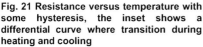

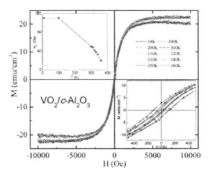

Fig. 21 Resistance versus temperature with some hysteresis, the inset shows a differential curve where transition during heating and cooling

Fig. 22 Magnetization (M) versus field (H) for VO2 films, showing a finite coercivity at room temperature

IX. NOVEL PEROVSKITE BASED SMART SENSORS

Neodymium nicklate ($NdNiO_3$) is a interesting material (Fig 23(a)) with a sharp transition in resistivity, where the transition temperature can be controlled via epitaxial strains (39-41). We have used the lattice-mismatch epitaxial strain, induced by the constraint of epitaxy, to tune the metal-insulator (M-I) transition temperature of $NdNiO_3$ films grown on Si(100) substrate. Films were integrated with the Si(100) substrate using several combinations of thin buffer layers. A systematic variation in the electrical transport properties has been observed with the change in the lattice mismatch between $NdNiO_3$ and the buffer layer just below it. It was shown that by the proper selection of the substrate and thickness of film, it is possible to control and precisely tune the M-I transition temperature of $NdNiO_3$ to any desired value between 12 and 300 K (as shown in Fig 23(d). Fig 23 (b) shows epitaxial integration on Si(100) through $TiN/MgO/SrTiO_3$ buffer layers, and the selected-area diffraction pattern in Fig 23 (c) shows epitaxial alignment of all the layers. Fine control over the M-I transition temperature of these films is likely to boost the potential of these films for their applications in bolometers, actuators, and thermal/optical switches in next-generation perovskite-based microelectronic devices.

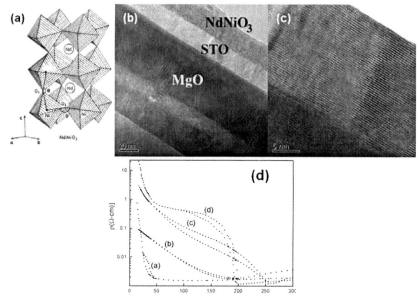

Fig. 23 (a) NdNiO$_3$ structure; (b) cross-section TEM showing epitaxial integration with Si(100) by using TiN, MgO and SrTiO$_3$ buffer layers; (c) high-resolution TEM and inset diffraction; and (d) resistance versus temperature under different strains

X. NANOTECHNOLOGY TO ENHANCE FUEL EFFICIENCY AND SAVE ENVIRONMENT

Finally, we show the impact of nanotechnology in improving the fuel efficiency and reducing the pollution (23). We show a substantial improvement in auto and truck fuel efficiency with concomitant reduction in wear and environmental pollutants. A novel concept in engine oil additives is proposed, where a combination of multifunctional nanoparticles (hard) and microparticles (layered soft) is added to the engine oil to smoothen and polish metallic surfaces and embed nanoparticles in the near surface regions, thereby reducing friction and wear by workhardening mechanisms (US Patent Pending, International Patent Granted). Inorganic ceramic nanoparticles of a certain range of sizes are created by pulsed laser ablation and dispersed in oil that can be added to the crank case oil of an automobile engine. Fig 24 shows improvement in fuel efficiency with the addition nano oil treatment. There is 20%

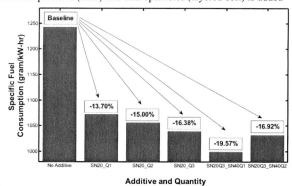

Fig. 24 Fuel efficiency with different treatments

improvement in fuel efficiency in stationary engines at the optimum dose. In the highway testing there is almost a factor of two enhancement up to 35-40% improvement because of the momentum based enhancement. Fig. 25 shows a drastic reduction in friction after the nano oil treatment, showing the basic mechanism improvement in fuel efficiency.

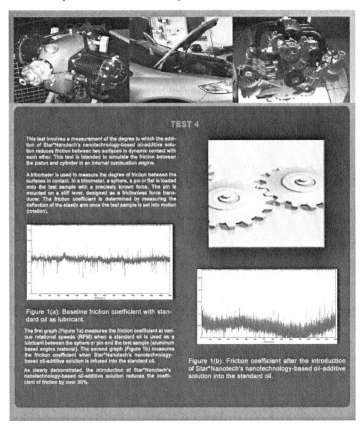

Fig. 25 Reduction in friction before and after nano oil treatment

SUMMARY AND CONCLUSIONS

This review addresses unprecedented opportunities and challenges in nanostructuring of solid state materials to improve their performance for the good of the society at large. Nanostructuring has a great potential in terms of obtaining the property of a perfect material, provided we can solve the problems associated with surfaces and interfaces of nanoscale materials. The transition from nanoscience to nanotechnology to society (through manufacturing of goods of improved performance) can be achieved through the systematic approach described in the transition octahedron. Some examples of this systematic approach and potential benefits, particularly in fabrication of light emitting diodes for solid state lighting and display devices, are discussed in this review.

Other topics include nanomagnetics, nanophotonics, smart sensors with integrated functionalities, and impact of nanotechnology on energy and environment. We specifically address nanosystems based upon nanodots and nanolayered materials synthesized by thin film deposition techniques, where recurring themes include nanostructuring of materials to improve performance; thin film epitaxy across the misfit scale for orientation controls; control of defects, interfaces and strains; and integration of nanoscale devices with (100) silicon based microelectronics and nanoelectronics. The control of structure, orientation, and composition is critical to obtaining novel properties for improved performance. The control of orientation and structure requires epitaxy across the misfit scale, which can be achieved through our recent invention of domain matching epitaxy. The DME is based upon matching of integral multiples of lattice planes across the film-substrate interface. The misfit in between the integral multiples can be accommodated by the principle of domain variation, where two domains alternate with a certain frequency to relax the misfit fully at the temperature of processing. This 22.5% percent lattice misfit (which is encountered in technologically important TiN/Si(100) system) can be accommodated by alternating 4/3 and 5/4 domains with equal frequency. The DME is critical for integration of different functionalities on Si(100), which is the substrate of choice for microelectronics and nanoelectronics, and other substrates with varying lattice misfits. The integration of novel perovskites and VO_2 based sensors on Si(100) and sapphire substrates is illustrated through the DME paradigm. Finally we cover the impact of nanotechnology on energy and environment through nano oil additives to improve fuel efficiency and reduce exhaust pollutants.

ACKNOWLEDGEMENT:
This research was supported by the National Science Foundation and parts by Army Research Office. The Author is very grateful to graduate students, postdoctral fellows and Professor John Prater for their immense contributions.

REFERENCES:

(1) R. P. Feynman, Annual Meeting of American Physical Society, 1959 [Caltech Engineering and Science journal, 4, 1(1960)].
(2) J. Narayan, Y. Chen, and R. M. Moon, Phys. Rev. Lett. 46, 149(1981); J. Narayan and Y. Chen, US Patent 4,376,755(1983).
(3) R. E. Smalley, MRS (Materials Research Society) Bulletin 30, 412 (2005).
(4) J. Narayan, "Critical Size for Defects in Nanostructured Materials," J. Appl. Phys. **100**, 034039 (2006).
(5) J. Narayan, "New Frontier in Thin Film Epitaxy and Nanostructured Materials," Int. J. Nanotech. 6, 493 (2009).
(6) J. Narayan and A. Tiwari, Journal of Nanoscience and Nanotechnology, 4, 724 (2004); US Patent # 2004/0119064 (June 24, 2004).
(7) J. Narayan, Met and Mat Trans B36, 5(2005)-2004 ASM Edward DeMille Campbell Lecture.
(8) V. A. Shchukin and D. Bimberg, Rev. of Mod. Phys. 71, 1125, 1999.
(9) C. J. brinker, Y. Lu, A. sellinger, and H. Fan, "Evaporation Induced Self Assembly," Advanced Materials 11, 579 (1999).
(10) L. Vayssieres, H. Hogfeldt, and S. E. Lindquist,"Purpose-built metal oxide materials, The emergence of next-generation smart materials," Pure Appl. Chem. 72, 47(2000).
(11) G. M. Whitesides and B. Grzybowski, "Self Assembly at ALL Scales," Science 295, 2418 (2002).
(12) L. Vayssieres, "New quantum confined visible light-active semiconductors," This volume
(13) Narayan, J., Tiwari, P., Chen, X., Singh, J., Chowdhury, R., and Zheleva, T., "Epitaxial growth of TiN films on (100) silicon substrate," Appl. Phys. Lett. **61**, 1290(1992); US patent 5,406,123 (1995).

(14)   J. Narayan and B. C. Larson, J. Appl. Phys. 93, 278(2003); US Patents on Domain Epitaxy # 5,406,123; 6,955,985.

(15)   J. Narayan, H. Wang, J. Ye, S. –J. Hon, K. Fox, J. C. Chen, H. K. Choi, and J. C. C. Fan, Effect of thickness variation in high-efficiency InGaN/GaN light-emitting diodes, Appl. Phys. Lett., 81, 841 (2002).

(16)   J. Narayan, Efficient light emitting diodes and lasers, US Patent 6,881,983 (2005) and   US Patent 6,847,052 (2005).

(17)   J. Narayan, H. Wang, T. –H. Oh, H. K. Choi, and J. C. C. Fan, Formation of epitaxial Au/Ni/Au ohmic contacts to p-GaN, Appl. Phys. Lett., 81, 3978 (2002).

(18)   J. Narayan, Electrode for p-type gallium nitride-based semiconductors, US Patent 6,734,091 (2004).

(19)   J. Narayan and V. M. Bhosle, J. Appl. Phys. 100, 103524 (2006).

(20)   A. Gupta A. Gupta , R. Aggarwal, P. Gupta, T. Dutta, Roger J. Narayan, and J. Narayan, Appl. Phys. Lett. 95, 111915 (2009).

(21)   T. Yang, C. Jin, R. Aggarwal, R. J. Narayan, and J. Narayan,   J. Mater. Res. (in press Mar. 2010)

(22)   T. Yang, S. Nori, H. Zhou and J. Narayan, Appl. Phys. Lett. 95, 102506 (2009).

(23)   J. Narayan,"Lubricant having nanoparticles and Microparticles to enhance fuel efficiency, and a laser synthesis method to create dispersed nanoparticles," US Patent #7,994,105 (2010).

(24)   J. Narayan, J. Appl. Phys. 37, 2703 (1985).

(25)   J.P. Hirth and J. Lothe, Theory of Dislocations, P. (1998) John Wiley, New York.

(26)   G. R. Trichy, J. Narayan, and H. Zhou, Appl. Phys. Lett., 89, 132502 (2006).

(27)   G. R. Trichy, D. Chakraborti, and J. Narayan, J. Appl. Phys., 102, 033901 (2007).

(28)   G. R. Trichy, D. Chakraborti, and J. Narayan, Appl. Phys. Lett., 92, 102504 (2008).

(29)   C. J. Humphreys, Solid-State Lighting, MRS Bull., 33, 459-477 (2008).

(30)   J. K. Kim and E. F. Schubert, Transcending the replacement paradigm of solid-state lighting, Optics Express, 16(26), 21835-21842 (2008).

(31)   Basic Research Needs for Solid-State Lighting, Report of the Basic Energy Sciences Workshop on Solid-State Lighting May 22-24, 2006. (http://www.er.doe.gov/production/bes/reports/files/SSL_rpt.pdf)

(32)   J. Narayan, P. Pant, A. Chugh, H. Choi, and J. C. C. Fan, Characteristics of nucleation layer and epitaxy in GaN/ sapphire heterostructures, J. Appl. Phys., 99, 054313 (2006).

(33)   S. C. Jain, M. Willander, J. Narayan, and R. van Overstraeten, III-nitrides, growth, characterization and properties, J. Appl. Phys. 87, 965 (2000).

(34)   M. Rao, D. Kim, and S. Mahajan, Compositional dependence of phase separation in InGaN layers, Appl. Phys. Lett., 85, 1961 (2004).

(35)   M. Rao, N. Newman, and S. Mahajan, The formation of ordered structures in InGaN layers, S. Scripta Mat., 56, 33 (2007).

(36)   X. Ni, Q. Fan, R. Shimada, Ü. Özgür, and H. Morkoç, Reduction of efficiency droop in InGaN light emitting diodes by coupled quantum wells, Appl. Phys. Lett., 93, 171113 (2008).

(37)   M. F. Schubert, J. Xu, J. K. Kim, E. F. Schubert, M. H. Kim, S. Yoon, S. M. Lee, C. Sone, T. Sakong, and Y. Park, Polarization-matched GaInN/AlGaInN multi-quantum-well light-emitting diodes with reduced efficiency droop, Appl. Phys. Lett., 93, 041102 (2008).

(38)   M. J. Galtrey, R. A. Oliver, M. J. Kappers, C. Humpherys, P. H. Clifton, D. Larson, D. W. Saxey, and A. Cerezo, J. Appl. Phys 104, 013524 (2008).

(39)   A. Tiwari, J. Narayan, and C. Jin,"Growth of epitaxial NdNiO$_3$ and integration with Si(100)," Appl. Phys Lett. 80, 1337 (2002).

(40)   A. Tiwari, C. Jin and J. Narayan," Strain-induced tuning of metal-insulator transition," Appl. Phys Lett. 80, 4039 (2002).

(41)   A. K. Sharma, J. Narayan, C. Jin,"Integration of PZT epilayers with Si(100) by domain epitaxy," Appl. Phys Lett. 76, 1458 (2000).

Thermal Management
Materials and Technologies

MEASUREMENT OF THERMAL CONDUCTIVITY OF GRAPHITIC FOAMS

Kevin Drummond and Khairul Alam[*]
Department of Mechanical Engineering, Ohio University, Athens, OH 45701, USA

ABSTRACT

Highly conductive graphitic foams are currently being studied for heat sink and thermal storage applications. This paper describes the experimental research at Ohio University to determine the thermal conductivity of graphitic carbon foam. The standard methods for determination of thermal conductivity generally work well for solid materials, but these methods may produce significant errors with graphitic foams because of convection and interface resistance. Therefore, a method has been developed in which the open celled graphitic foam is infiltrated by a low conductivity polymer. The resulting composite is a dense solid that can be measured accurately by classical methods. It is shown that the presence of the polymer has negligible effect on the thermal conductivity of the graphitic foam; therefore, the bulk conductivity of the foam is reasonably well approximated by the conductivity of the infiltrated foam.

INTRODUCTION

Open cell metal foams have been used in thermal energy storage (TES) systems and heat sinks for many years. Recently, graphitic foams have emerged as an alternative to metal foams in these applications. Graphitic foams have certain advantages over traditional fins and metal foams; these include high specific surface area, variable pore size, inertness to chemicals, and potential for high thermal conductivity[1]. The high surface area due to the cellular geometry of foams makes them good candidates for heat transfer to a fluid or to a phase change material (PCM).

Significant research is being conducted to develop new graphitic foams, and to determine their properties and their performance in heat sinks and TES systems. Stansberry and Singer[2] developed a method of creating highly graphitic carbon foams and have experimentally measured thermal conductivities in excess of 150 W/mK. Klett et al.[3] have reported thermal conductivities up to 180 W/mK for graphitic foams produced at Oak Ridge National Laboratories (ORNL). This is an order of magnitude higher than metal foams of comparable porosity[4,5].

An accurate determination of the thermal conductivity of a porous medium can be a difficult task when the conductivity values are extremely high. The classical guarded hotplate method (ASTM C177) is specified to be accurate only up to about 10 W/mK[6]. The longitudinal comparative heat flow technique (ASTM E1225) is useful for materials with thermal conductivity up to 200 W/mK[7]. However, this method is specified for homogeneous solids, and can give erroneous results if porous samples are used. Therefore alternate methods, such as the flash diffusivity method (ASTM E1461)[8], are often selected. The flash diffusivity technique measures the diffusivity of the sample by using a laser or a xenon flash to heat up one surface of a sample and then measuring the time history of the temperature rise on the opposite face of the sample. The thermal conductivity is then calculated by the following formula:

$$k = \alpha \rho c_p \qquad (1)$$

where k is thermal conductivity, α is thermal diffusivity, ρ is density, and c_p is specific heat.

---

[*] To whom correspondence should be addressed

185

It should be noted the flash diffusivity technique is an indirect method to measure the thermal conductivity and is also designed for solid materials[9]. However, the flash diffusivity method can be reasonably accurate if the pores are small compared to the overall sample thickness so that the energy pulse of the flash does not penetrate the sample significantly. Because the laser flash method result is proportional to the square of the sample length[10], it can be expected that a partial penetration of the flash into the foam will reduce the effective length and overestimate the diffusivity. Therefore, the flash diffusivity method is much more appropriate for cellular graphitic foams with small pores than low density metal foams having large pores. As a result, this has been a popular method for determining the thermal diffusivity and conductivity of graphitic foams.

The objective of the present study is to develop accurate, direct thermal conductivity measurements of highly conductive graphitic foams. This is achieved by modifying the LaserComp FOX50 "Heat Flow Meter" instrument (based on ASTM C518) to eliminate the effect of interface resistance. The direct results are then compared with data from the indirect flash diffusivity method. Diffusivity data was obtained using the Anter Flashline 3000 instrument[11], described in the next section.

METHOD

Flash Diffusivity Method

The Anter Flashline 3000 instrument uses the laser flash technique to determine thermal diffusivity. A short pulse of light is exposed to the top surface of a sample. Heat is absorbed by the sample and the subsequent temperature rise is measured on the bottom surface. An ideal response is shown in Figure 1, below. Also shown is the half-time $(t_{1/2})$ which is defined as the time it takes to reach half of the maximum temperature. With the thickness of the sample (Δx) known, thermal diffusivity is calculated using Equation 2[10].

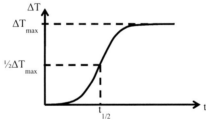

Figure 1. Ideal temperature response from laser flash technique

$$\alpha = 1.388\frac{(\Delta x)^2}{t_{1/2}} \qquad (2)$$

It is worth noting that the maximum temperature is not measured, only the time to reach half of the maximum. Therefore the light intensity can be changed without affecting the results. Also, the instrument used in this study accepts samples up to 1" thick. This is considerably larger than the pore sizes of the foams tested, decreasing the effect of laser penetration on thermal diffusivity measurement.

Heat Flow Meter

The FOX50 instrument[12] uses a modified version of Fourier's law of conduction, shown in Equation 4, to calculate thermal conductivity. In Equations 3 and 4, q is heat flux, ΔT is temperature gradient across sample, Δx is thickness of the sample, k is thermal conductivity, and R_s is surface contact resistance.

$$q = -\frac{\Delta T}{\left(\frac{\Delta x}{k}\right)} \tag{3}$$

$$q = -\frac{\Delta T}{\left(\frac{\Delta x}{k}\right) + 2R_s} \tag{4}$$

The assumptions made for Equation 4 to be valid are 1) conduction is the only mode of heat transfer, 2) the measured temperature gradient is the same as that in the sample, 3) consistent cross-sectional heat flux across the area of the sample, and 4) contact resistance of the sample being tested is the same as the calibration sample.

The instrument consists of top and bottom plates which are independently heated and cooled between -10°C and 110°C. A heat flux meter and thermocouple are located near the surface of each plate, respectively. The instrument works by creating a steady-state temperature gradient across the sample, then measuring the heat flux in and out, top and bottom temperatures, and thickness of the sample. Traditionally, the temperatures are measured inside the top and bottom plates rather than inside the sample. This can cause significant errors if surface contact resistance is not properly accounted for, as seen in Figure 2. The temperature jump present at both surfaces is caused by surface contact resistance, which occurs due to the misalignment of atomic planes and microscopic air gaps present on each surface. Surface contact resistance values are found by testing samples of known conductivity. These predetermined values are then used while testing samples of unknown conductivity. While this method is accurate for samples with low thermal conductivity (<10 W/mK), significant errors arise when dealing with highly conductive materials, as is the case with graphitic foams.

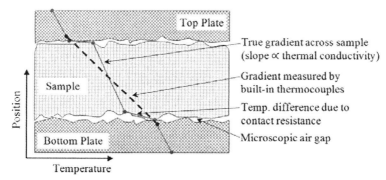

Figure 2. Temperature gradient created by guarded heat flow instrument

Both the heat flux meters and thermocouples require a solid surface and consistent cross-sectional heat flux to function properly. The foams tested in this study have porosities between 75% and 95%. Therefore, on average, this amount of the plate would not be in contact with solid foam if left unfilled. To create a smooth surface, epoxy is infiltrated into the pores of the foam then machined down and polished. Infiltration was done under a vacuum and low viscosity resin was used to help ensure complete infiltration. Since the resin has a comparatively low conductivity, it has a negligible effect on the overall sample conductivity. Infiltration also eliminates the effects of free convection during testing.

The FOX50 instrument is designed to test materials with thermal conductivity between 0.1 W/mK and 10 W/mK[12]. Since the samples tested in this research are much higher conductivity—up to 150 W/mK—special methods are used to obtain more accurate data. First, thermal paste is used to fill the air gap between the sample and the top and bottom plates. This creates a more uniform heat flux through the sample by eliminating areas of high thermal resistance along the surfaces; it also decreases the overall thermal resistance across the surface. Also, the sample's temperature gradient was measured by drilling small radial holes near the top and bottom surfaces of the samples and inserting thermocouples; a schematic of the test samples is shown in Figure 3 below. Using inserted thermocouples provided temperature readings in the sample rather than on the plate surface resulting in a more accurate measurement of thermal conductivity.

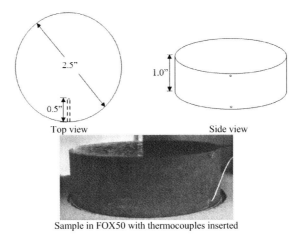

Top view   Side view

Sample in FOX50 with thermocouples inserted

Figure 3. Locations of thermocouple holes

RESULTS

Tests were conducted on graphitic foams to compare with the conductivity values from the direct method and the flash diffusivity method. For this study, the samples were made from two densities of foam—0.10 g/cc and 0.22 g/cc. In graphitic foams, the conductivity increases with the density of the foam. In this case, the lower density foam has a thermal conductivity between 10 and 15 W/mK while the higher density foam has a thermal conductivity between 45 and 50 W/mK. It should be noted that these materials exhibit anisotropic behavior. The thermal conductivity ranges listed here are associated with the foaming or rise direction. Typically, the transverse values are considerably less (approximately, 50%). All the measurements in this study were taken in the rise direction.

A single disk sample, 2.5" in diameter (Figure 2) was cut from a block of the lower density foam for thermal conductivity tests in the FOX50 instrument. Several 1" cubes (4 to 5) were cut from the same block of the foam. The cubes were tested in the flash diffusivity instrument–with each sample measured 3 times and the average diffusivity of the three shots shown in Table I. It should be noted that the diffusivity of each sample did not vary significantly with repeated runs. The variation from sample to sample is the dominant factor in the standard deviation of the results. Five samples were tested and the average of the results was taken to be representative of the block of the source material, and thus representative of the disk. The average thermal diffusivity was found to be 1.864 cm^2/s, with a standard deviation of 0.115 cm^2/s. The thermal conductivity, calculated using Equation 1, was calculated to be 13.13 W/mK with a standard deviation of 0.81 W/mK. The relative standard deviation of the thermal diffusivity and derived conductivity values is 6.1%.

Table I. Diffusivity results for lower density foam (density=0.10 g/cc)

Sample #	Average Measured Thermal Diffusivity (cm^2/s)	Derived Thermal Conductivity (W/mK)
1	1.9849	14.09
2	1.8841	13.38
3	1.7188	12.20
4	1.7724	12.58
5	1.9593	13.91
Average	1.8639	13.23
Std Dev	0.115	0.81

As described earlier, the disk was infiltrated with epoxy, coated with a thermal paste on the top and the bottom surfaces and runs were carried out to measure its thermal conductivity. The thermal conductivity results of the single disc sample shows variation from run to run because of the variation in the thermal interface produced by the thermal paste, and of the variability within the disk as seen by the heat flow meters in the FOX50 instrument. The runs were repeated until the standard deviation of the measured thermal conductivity results was comparable to or less than the standard deviation of the thermal diffusivity results. The average thermal conductivity was determined to be 14.12 W/mK, with a standard deviation of 0.57 W/mK (4.0% relative standard deviation). The average measured value is approximately 6% higher than the average of the derived thermal conductivity results.

Table II. Measured thermal conductivity of lower density foam (density=0.10 g/cc)

Run #	Measured Thermal Conductivity (W/mK)
1	13.95
2	13.50
3	14.79
4	14.65
5	13.71
Average	14.12
Std Dev	0.57

The above measurements were then repeated by using samples cut out from higher density foam block (0.22 g/cc). The results are shown below in Tables III and IV. The thermal conductivity derived from thermal diffusivity measurement is 48.66 W/mK with a standard deviation of 1.24 W/mK (2.5% relative standard deviation).

Table III. Diffusivity Results for higher density foam (density = 0.22 g/cc)

Sample #	Average Measured Thermal Diffusivity (cm²/s)	Derived Thermal Conductivity (W/mK)
1	3.1757	50.05
2	3.1739	50.03
3	3.0446	47.99
4	3.0616	48.26
5	2.9796	46.96
Average	3.0871	48.66
Std Dev	0.079	1.24

The measured thermal conductivity, as shown in Table IV, is 45.55 W/mK with a standard deviation of 1.43 W/mK (3.1% relative standard deviation). The difference between the average values of derived and measured thermal conductivity is approximately 7%, which is similar to the results for the lower density foam. However, it should be noted that the measured value is higher than the derived value of the lower conductivity foam, but the reverse is true with the higher conductivity foam.

Table IV. Measured thermal conductivity of higher density foam (density=0.22 g/cc)

Run #	Measured Thermal Conductivity (W/mK)
1	46.16
2	44.01
3	44.80
4	47.24
Average	45.55
Std Dev	1.43

DISCUSSION AND CONCLUSIONS

In this study, a method was developed to determine the thermal conductivity of high conductivity foams by modifying the FOX50 instrument, which uses the heat flow meter technique based on ASTM C518. The results are then compared with the thermal conductivity derived from thermal diffusivity measurements. The results show that the modified heat flow technique used in this study can be used to measure foams of high conductivity (up to 50 W/mK) within reasonable accuracy.

The thermal conductivity values of the foam samples measured by the two different methods showed standard deviation of up to 3.1% within the block of a higher conductivity (higher density) foam, and 6.1% in the lower conductivity (lower density) foam block. Variation in conductive graphitic foams can be expected, and is typically due to the variation in processing conditions within a furnace. To avoid the errors associated with such variations within the sample, the comparison of the two measurement methods should be conducted on the identical sample material. However, this was not feasible in this study due to the differences in the sample geometry in the two instruments. Future tests are planned to measure identical samples by direct and indirect methods to make a more accurate comparison of the two methods.

ACKNOWLEDGEMENTS

The authors would like to acknowledge the support provided by the Air Force Research Laboratory (AFRL, Dayton, OH), GrafTech International (Parma, OH, USA), and the Ohio Aerospace Institute (OAI, Cleveland, OH).

REFERENCES

1. Gallego, N. C. & Klett, J. W. Carbon foams for thermal management. *Carbon* **41**, 1461 (2003).
2. Stansberry, P. & Singer, L. Enhanced directional conductivity of graphitizable foam. 5 (2008).
3. Klett, J. W., McMillan, A. D., Gallego, N. C. & Walls, C. A. The role of structure on the thermal properties of graphitic foams. *Journal of Materials Science* **39**, 3659–3676 (2004).
4. Calmidi, V. V. & Mahajan, R. L. The effective thermal conductivity of high porosity fibrous metal foams. *Journal of Heat Transfer* **121**, (1999).
5. Salimon, A., Brechet, Y., Ashby, M. F. & Greer, A. L. Potential applications for steel and titanium metal foams. *Journal of Materials Science* **40**, 5793–5799 (2005).
6. ASTM Standard C177, 2010, "Standard Test Method for Steady-State Thermal Transmission Properties by Means of the Heat Flow Meter Apparatus," ASTM International, West Conshohocken, PA, 2010, DOI: 10.1520/C0177-10, www.astm.org.
7. ASTM Standard E1225, 2009, "Standard Test Method for Thermal Conductivity of Solids by Means of the Guarded-Comparative-Longitudinal Heat Flow Technique," ASTM International, West Conshohocken, PA, 2009, DOI: 10.1520/E1225-09, www.astm.org..
8. ASTM Standard E1461, 2009, "Standard Test Method for Thermal Diffusivity by the Flash Method," ASTM International, West Conshohocken, PA, 2007, DOI: 10.1520/E1461-07, www.astm.org.
9. Parker, W. J., Jenkins, R. J., Butler, C. P. & Abbott, G. L. Flash method of determining thermal diffusivity, heat capacity, and thermal conductivity. *Journal of Applied Physics* **32**, 1679–1684 (1961).
10. Taylor, R. E. Heat-pulse thermal diffusivity measurements. *High Temperature- High Pressure* **11**, 43–58 (1978).
11. Flashline 3000 Thermal Properties Analyzer. *Anter Corporation* (2007). <http://anter.com/FL3000.htm>
12. Statistics Sheet for FOX50. *LaserComp, Inc.* (2003). <http://www.lasercomp.com/fox50.htm>

EXAMINATION OF THE INTERCONNECTIVITY OF SiC IN A Si:SiC COMPOSITE SYSTEM

A. L. Marshall
M Cubed Technologies, Inc.
1 Tralee Industrial Park
Newark, DE 19711

ABSTRACT

Composites of silicon carbide particles (SiC_p) and silicon (Si) are fabricated by the reactive infiltration of molten Si into preforms of said particles and carbon. Depending upon the application, these materials can be used in many situations due to their favorable properties including high hardness, low thermal expansion, high thermal conductivity and high stiffness. Through heat treatment methods, necking of the SiC_p is promoted in the pre-infiltration state which will allow further tailoring of material properties. This study examines the effects of firing temperature on density, Young's modulus, and the interconnectivity of the SiC_p. Interconnectivity is examined through optical microscopy methods.

INTRODUCTION

Reaction Bonded SiC (RBSC) lends itself to many potential applications including: thermal applications due to the high thermal conductivity, low tailorable thermal expansion, and high specific stiffness; armor applications due to the high hardness, high Young's modulus, and low density; and also high-temperature wear applications due to the high hardness and stability at elevated temperatures. RBSC is a composite material made up of Si and SiC. Some of the properties are easily tailorable to specific applications depending upon the starting materials used.

The basic process for making a RBSC part involves four steps. First, one must make a slurry of SiC, a carbon based binder, and water. Second, a preform is prepared through any number of casting or molding techniques, including: sedimentation, injection molding, filter casting, slip casting, etc. Third, the binder in the preform is converted to carbon by heating up the part in a nitrogen rich environment to the cracking temperature of the particulate binder. Last, the part is placed in a vacuum furnace in contact with a Si alloy and heated to a temperature under high vacuum such that the alloy turns molten and wicks into the preform and reacts with the carbon therein.[1,2] The residual Si fills the interstices, forming a fully dense composite with minimal dimensional change. A cartoon is provided below in Figure 1 highlighting this process.

High-precision applications have very demanding requirements for RBSC materials including good thermal stability, high specific stiffness, tight flatness tolerances, as well as a low wear rate. The motivation behind this particular study involves improving the thermal conductivity and thereby increasing the thermal stability as well as decreasing the amount of material loss due to wear through repeated use. To increase the thermal stability, two possible options are: substitute a material with a higher thermal conductivity or improve the thermal transfer by reducing the thermal barrier transfer. With the other requirement being to reduce the wear rate, the focus herein will be on reducing the thermal barrier by firing the parts in their preform state at various temperatures, and subsequently infiltrating them with molten Si. This will promote necking of the particles and provide an improved path for thermal transfer within the composite.

Typically, one could improve these properties by realizing that thermal conductivity increases with increasing grain size and that material wear loss reduces with increasing grain size. Thermal conductivity data is presented in Figure 2 to demonstrate this. Unfortunately, the ability to machine high-precision parts also decreases with increasing grain size as the grain size becomes relevant to the dimensional tolerancing needed.

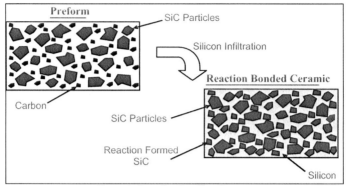

Figure 1. Reaction Bonding Cartoon

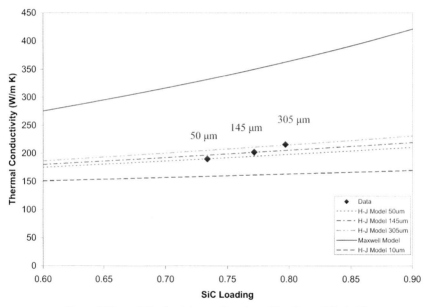

Figure 2. Thermal Conductivity as a Function of Loading and Grain Size
(particle size indicated above data points)

The Maxwell[3] model and Hasselman-Johnson[4,5] models were used to produce the curves provided in Figure 2. The equations are respectively provided below for reference:

$$\lambda_c = \lambda_m \left(\frac{2(1-V_p) + (\lambda_p/\lambda_m)(1+2V_p)}{(2+V_p) + (\lambda_p/\lambda_m)(1-V_p)} \right) \tag{1}$$

$$\lambda_c = \lambda_m \left(\frac{2(1-V_p) + (\lambda_p/\lambda_m)(1+2V_p) + 4(\lambda_p/Dh)(1-V_p)}{(2+V_p) + (\lambda_p/\lambda_m)(1-V_p) + 2(\lambda_p/Dh)(2+V_p)} \right) \tag{2}$$

here, D represents the reinforcement particle diameter, and h is the thermal boundary conductance. The thermal boundary conductance was estimated based on the data points in Figure 2. V is the volume fraction; λ is the thermal conductivity; and c, m, and p stand for composite, matrix, and particulate, respectively. The thermal boundary conductance is dependant on the contact between the faces along with the CTE mismatch. When the particle diameter becomes large (D→∞) Equation 2 reduces to the Maxwell model, Equation 1. This is the main goal of the pre-infiltration firing; increased necking will allow these materials to behave as if the diameter is larger than its initial state. This should also decrease material wear as the particles will be more tightly bound to each other. An optimal connectivity will be one that improves these properties without appreciably reducing machinablity.

EXPERIMENTAL PROCEDURE
    Two sets of samples were produced as preforms, nominally 10 μm SiC_p preforms and bimodal 50 and 10 μm SiC_p preforms. Preforms were raised to temperature in a partial nitrogen atmosphere at a rate of 8°C/min. Preforms were held at temperature for one hour at 1500°C, 1800°C, 2100°C and two hours at 2150°C and 2250°C. The longer time at higher temperature was performed to promote recrystallization of the SiC_p. An optical microscope was used to capture images of each microstructure and ImageJ software was utilized to perform the image analysis. Density and Young's modulus were also taken for completeness.

RESULTS & DISCUSSION
    Density and Young's modulus data is presented in Figure 3. For the most part, the density and modulus did not change appreciably with respect to firing temperature. This was expected as the volume of the samples did not have a noticeable change upon firing. The 10 μm SiC_p did show a decrease in Young's modulus at the higher temperatures. With the density being indicative of loading, it is not currently understood why this occurred. As will be shown in the microstructures, complete infiltration occurred; residual porosity was not present as the pores were not closed during the firing.
    The microstructures of the samples are also provided in Figures 4 thorugh 7. Figures 4 and 5 contain microstructures for the 10 μm SiC_p, and Figures 6 and 7 contain the microstructures for the bimodal SiC_p samples. The light gray phase is the Si and the darker gray phase represents the SiC. When examining the images one can qualitatively perceive an increase in the interconnectivity or necking of the SiC_p. This can be determined by examining the amount of isolated SiC_p chains and how the number of these chains reduces with the higher temperatures.

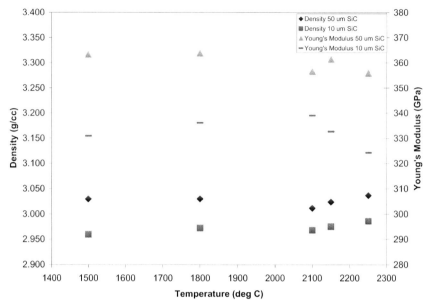

Figure 3. Density and Young's Modulus vs. Temperature

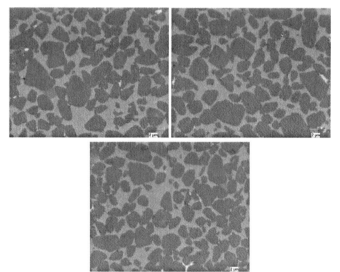

Figure 4. 10 μm SiC_p Micostructures: (left to right, 1500°C, 1800°C, and 2100°C samples)

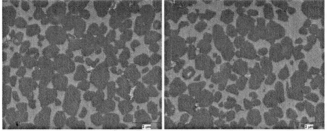

Figure 5. 10 μm SiC$_p$ Micostructures: (left to right, 2150°C and 2250°C samples)

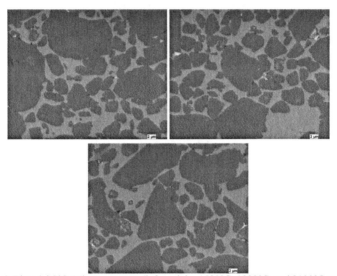

Figure 6. Bimodal SiC$_p$ Micostructures: (left to right, 1500°C, 1800°C, and 2100°C samples)

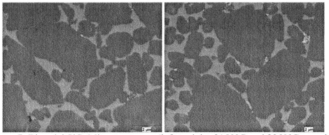

Figure 7. Bimodal SiC$_p$ Micostructures: (left to right, 2150°C and 2250°C samples)

Image analysis was also performed to provide a quantitative assessment of the interconnectivity and how it changes with firing temperature. This was performed on the 10 μm SiC_p with the result being representative of the bimodal samples as well. Ten images were taken at 1000x magnification for each firing temperature. These images were taken at random and are construed to be representative of the whole sample. Each individual image represented an area of approximately 8825 $μm^2$. As these materials are homogeneous on a macro level, the cross sections are representative of the whole volume.

As previously stated, ImageJ software was utilized during the analysis. The "Analyze Particles" feature was used to count the number of isolated regions in each image. The data was filtered by excluding any data which represented an area less than 0.196 $μm^2$. This represents the cross sectional area of a 500 nm spherical particle; anything with an area below this was determined to be extraneous as the distribution for the 10 μm SiC_p material is normal around 10 μm. A chart is presented in Figure 8 with the data.

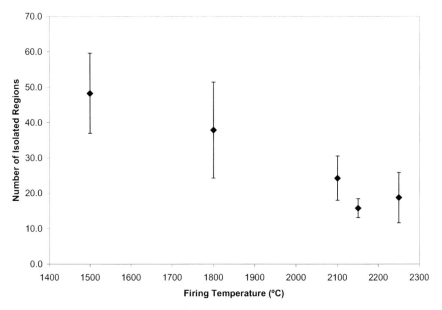

Figure 8. Number of Isolated Regions vs. Firing Temperature

As expected, the number of isolated regions decreases with increasing firing temperature indicating increasing interconnectivity or necking of the SiC_p. This supports the qualitative understanding one gleans by examining the images presented in Figures 4 though 7. Ignoring the slight difference in firing time, the interconnectivity appears to have a linear dependence on firing temperature. This should lead to a decrease in the thermal barrier resistance inherent in the material as the surface area should be decreased. In other words, the effective diameter should have been increased leading the material to behave more like the Maxwell model than the Hasselman-Johnson model, Equations 1 and 2 respectively.

SUMMARY

Various pre-infiltration firing temperatures were examined to improve the interconnectivity of the SiC_p in two RBSC materials. Successful, pore-free, structures were subsequently produced with typical reaction bonding production methods. A qualitative, microstructures, and quantitative, image analysis, examination was performed to understand how the firing temperatures affect interconnectivity. At the prescribed conditions, a linear relationship is noted between the number of connected regions and the firing temperature in a given representative area.

REFERENCES

[1] M. K. Aghajanian, B. N. Morgan, J. R. Singh, J. Mears, R. A. Wolffe, A New Family of Reaction Bonded Ceramics for Armor Applications, *Ceramic Transactions*, **134**, J. W. McCauley et al. editors, 527-40 (2002).

[2] P. G. Karandikar, M. K. Aghajanian and B. N. Morgan, Complex, Net-Shape Ceramic Composite Components for Structural, Lithography, Mirror and Armor Applications, *Ceram. Eng. Sci. Proc.*, **24** [4] 561-6 (2003).

[3] Z. Hashin, S. Shtrikman, "A Variational Approach to the Theory of the Elastic Behaviour of Multiphase Materials," *J. Mech. Phys. Solids*, **11** 127-140 (1963).

[4] L. Zhang, et. Al., "Thermo-physical and mechanical properties of high volume fraction SiC_p/Cu composites prepared by pressureless infiltration," *Mat. Sci. & Eng.* **A 489** 285-293 (2008).

[5] A.G. Every, et. al., "The Effect of Particle Size on the Thermal Conductivity of ZnS/Diamond Composites," Acta metal. Mater. Vol 40, No 1. 123-129 (1992).

Author Index